ISBN 978-1-331-51826-6
PIBN 10200828

1 MONTH OF
FREE
READING

at

www.ForgottenBooks.com

By purchasing this book you are eligible for one month membership to ForgottenBooks.com, giving you unlimited access to our entire collection of over 700,000 titles via our web site and mobile apps.

To claim your free month visit:

www.forgottenbooks.com/free200828

Practical Mechanics and Allied Subjects
by Joseph W. L. Hale

An Introduction to the Study of Electrical Engineering
by Henry Hutchison Norris

Mechanical Engineering for Beginners
by R. S. M'Laren

Lathe Work for Beginners
A Practical Treatise, by Raymond Francis Yates

Sanitary Engineering
by William Paul Gerhard

A Text-Book of Mechanical Drawing and Elementary Machine Design
by John S. Reid

Measuring Tools
by Unknown Author

The Design of Simple Roof-Trusses in Wood and Steel
With an Introduction to the Elements of Graphic Statics, by Malverd Abijah Howe

The Wonder Book of Engineering Wonders
by Harry Golding

Plain and Reinforced Concrete Hand Book
by William O. Lichtner

Principles and Design of Aëroplanes
by Herbert Chatley

The Practical Tool-Maker and Designer
by Herbert S. Wilson

Thread-Cutting Methods
by Franklin Day Jones

A Treatise on Lathes and Turning
Simple, Mechanical, and Ornamental, by W. Henry Northcott

Water-Supply Engineering
The Designing and Constructing of Water-Supply Systems, by Amory Prescott Folwell

The Strength of Materials
by J. A. Ewing

Practical Safety Methods and Devices, Manufacturing and Engineering
by George Alvin Cowee

Shop Practice for Home Mechanics
Use of Tools, Shop Processes, Construction of Small Machines, by Raymond Francis Yates

Skeleton Structures
Especially in Their Application to the Building of Steel and Iron Bridges, by Olaus Henrici

Steam Boilers
Design, Construction, and Management, by William Henry Shock

GRIFFITHS' GUIDE

TO THE

IRON TRADE OF GREAT BRITAIN

WITH PLATES AND ILLUSTRATIONS

CONTAINS

An Elaborate Review of the Iron & Coal Trades for Last Year

ADDRESSES AND NAMES OF ALL IRONMASTERS

WITH A LIST OF BLAST FURNACES, IRON MANUFACTORIES,
AND OTHER STATISTICS AND INFORMATION
RESPECTING IRON AND COAL

which may be useful to

MERCHANTS		COALOWNERS
	BROKERS	
BANKERS		IRONMASTERS

AND ALL OTHERS INTERESTED IN THE IRON TRADE

BY

SAMUEL GRIFFITHS

Editor of 'The London Iron Trade Exchange'

LONDON
PUBLISHED FOR THE PROPRIETOR, 133 CANNON STREET
1873

Price One Guinea

DEDICATION.

TO

Sir Sydney Hedley Waterlow, Baronet,

LORD MAYOR OF LONDON.

My Lord,

Although your Lordship is not directly connected with the Iron Trade, the vast importance of its interests to the prosperity of these Kingdoms is well understood by your Lordship, and must be my apology for availing myself of your sanction to dedicate ' The Guide to the Iron Trade of Great Britain ' to the Chief Magistrate of the first commercial City in the world.

Although the Coal and Iron Mines and Furnaces are fixed at a considerable distance from your Lordship's jurisdiction, the capital necessary for their development, and the merchant princes who serve the craving nations of the earth with Iron, are mainly located in this great Metropolis. Without the Bank of England, Messrs. Glyn's, and other establishments of a kindred nature, it would be difficult to arrange the great monetary exchanges required, and, indeed, absolutely necessary for firms engaged in the Coal and Iron Trades with foreign countries.

We are indebted in no small degree for the gigantic proportions which these trades have assumed in the civilised world to the banking facilities afforded by, and the known probity and high honour of the Bankers and Merchants and Brokers of the City, which, of its own free will, called your Lordship to preside over its ancient municipality, and dispense, as yourself and predecessors have done, the civic hospitality to the mightiest potentates of the earth.

Permit me, my Lord, to offer a fervent prayer that your life and health may long be spared, and that you may continue to act under the empire of the motto on the façade of the Royal Exchange, which I also acknowledge—' The earth is the Lord's and the fulness thereof,' and to attribute in the very highest degree our prosperity as a nation to our belief in these inspired words.

I remain, with profound respect, Your Lordship's

Most obliged and humble Servant,

SAMUEL GRIFFITHS.

84, CANNON STREET.

PREFACE.

THE pages of the 'Guide to the Iron Trade of Great Britain' have been inspired under the impulse of a strong desire to afford practical information to the London and Liverpool merchants whose business operations bring them into frequent contact with the producers of the most important staple manufactures of the United Kingdom. The author is well aware that the task might have been performed by others more capable of doing justice to it than himself. However, as the necessity of a Guide to the Trade at this juncture is admitted on all hands, he has with diffidence undertaken the responsibility; and craves indulgence and forbearance from those who may be disposed to cavil with the Guide, which is not as perfect in detail as the author could wish, notwithstanding the assiduous labours which have been bestowed upon it. The great object of the author in this work has been to exhibit the grandeur of our great staple, and by the accumulation of patent facts, show the important position we occupy throughout the world in the manufacture of Iron and other metallurgical industries. Should this

first effort be well received, every exertion will be made
to render the Second Edition more perfect and accept-
able. The work, as it now stands, is offered to a
discerning public, by whose verdict the author will
be content to abide.

S. G.

INTRODUCTION.

I FEEL that an apology is justly due from me to the Subscribers to the 'Guide to the Iron Trade of Great Britain' in consequence of the delay which has occurred in the publication since its advent was first announced. Deeply as I regretted this shortcoming, notwithstanding my untiring efforts, I am grieved to say that unavoidable difficulties, and obstacles insurmountable and beyond my control, prevented an earlier publication. In the first place, the returns necessary to complete the statistics were delayed, perhaps not willingly, in various quarters. In the second, the plates and engravings which it was from the first intended should accompany the work, were often delayed, and frequently required to be abandoned for more truthful and correct representations. In the third place, the editorial labour exacted from me by my old 'Iron Trade Circular,' now called the 'Iron Trade Exchange,' has increased fiftyfold this year, which made imperative demands on my time, and large inroads on the leisure hours which otherwise would have been willingly consecrated to the 'Guide.' In the fourth place, as the 'Iron Trade Exchange' is emphatically a subscription newspaper, and all subscribers thereto are entitled by their subscriptions to receive answers to all letters demanding information on any subject in respect to the Iron Trade, my personal attention in dictating replies to these applications, particularly when the Foreign Mails came in, was and continues to be absolutely necessary. These pressing engagements, with my general business as a merchant and metal agent, so thoroughly occupied my time as to render the task of completing my book difficult; and I can with truth say that the above causes and obstacles are entirely accountable for the delay, which has often vexed and annoyed me.

It formed no part of my original intention to write a scientific essay on the manufacture of Iron and Steel. I have therefore adhered from first to last, in the pages of the 'Guide,' to my original programme in this respect. A book of this character would of necessity occupy two or three volumes; and as Mr. Lowthian Bell, Mr. Henry Bessemer, and Sir William Fairbairn, have so ably served the Trade in this respect, I have confined myself more to short histories, practical hints and sketches, and patent facts and statistics in connection with the Trade and the art of Iron-making ; which, while they do not involve any particular *modus operandi*, may be interesting and, I hope, beneficial to Merchants and Ironmasters in all departments of the Trade.

Anyhow, this has been my design and object in the pages and illustrations which I offer with no small degree of diffidence to the Trade and the public.

The Review of the Iron Trade for 1872 has been written with care. The statistical facts embodied therein the author hopes may be useful as a reference. The Commercial Review for the same year will be found consistent with the facts developed in the extraordinary volume of demand which created such a revolution in prices as will stand for ever as the crowning incident of the Trade for 1872. All these leading facts I have, to the best of my knowledge, truthfully recorded.

The grandeur of our position as an Iron-making country has always been present to my mind; and although the 'Guide' frequently dilates with pride on our unrivalled produce of Iron and metallurgical industries, I have endeavoured to avoid comparisons with other countries which would have raised our Iron-masters in their own estimation at the expense of foreign competitors. I firmly believe that the enterprising leaders of the great industries of all nations are being, as it were, involuntarily drawn nearer together in fraternal ties, and that as civilisation progresses the interests of all countries will be found to be more identical and prosperous all round in proportion as brotherly feeling is cultivated, ministering continually, as it will, by interchanges of thought and good offices, and thus inspiring us constantly to instruct each other in the perfection of art, which is the handmaid to progress, civilisation, and social happiness. My statistics in regard to the blast furnaces will be found in their proper place, and the quantity of Iron made in 1872 not far off the mark.

With regard to my statistics of the puddling-furnaces, they are as perfect as they could be made under present circumstances. The revolution now going on in puddling is more general than those confined to one works in one locality would expect or believe. Numbers of Danks' and Casson's furnaces have already replaced the old furnace, and others of this class are in rapid course of erection.

This, with other circumstances, has rendered a list of the puddling-furnaces, at work and standing, perfect and complete, impossible. At the same time I hope my statistics in this respect will be considered satisfactory. I have to acknowledge with gratitude statistical and other information willingly afforded by Mr. Robert Hunt, her Majesty's custodian of mining records at the School of Mines. I am likewise indebted to Mr. Henry Bessemer for information personally given to me in regard to his wonderful process, which we all admire for the beneficial revolution it has effected in the Trade. I must also acknowledge my indebtedness to my late friend, Mr. Beecroft's book, from which I obtained valuable information in regard to mensuration, weights, and strength of materials, &c.

I am under the same obligations in regard to Scotch Iron to Messrs. Feldtman, the old Iron brokers of Glasgow; and for ancient statistics in regard to Scotch Iron both in respect to price and qualities I am indebted to my old friend Thomas Thorburn, of Glasgow.

I have heartily to thank the Ironmasters and Manufacturers of England and Scotland for the polite manner in which the portals of their establishments have always been opened when my visits, which have been numerous, have been made, and hope that my reports of anything I may have seen have been consistent with discretion, in sometimes suppressing the publication of new discoveries in machinery which, from the just claims of private interest, ought not to be divulged. I may, without fear of contradiction, say that I have visited as many Iron Works and metallurgical establishments in this and other countries as any man living, and never was refused admission but once in my life. Only the other day I was permitted to look over the Patent Nut and Bolt Company's London Works, where, according to my view, the greatest wonders in England are performed by their patent machines. Indeed so marvellous were the results of the working of these machines, that I could not have believed it possible from any other evidence than my own eyes.

In conclusion, therefore, I must be permitted to

acknowledge with gratitude the kind and polite attention I have invariably received in these numerous
visits of general observation at our Iron-making and
great metallurgical establishments, which perform so
important a part in their ministrations to the wealth
and stability of these kingdoms.

1. OBSERVATIONS ON THE QUALITY OF IRON.

THE quantities of all other metals consumed, when compared with the consumption of Iron, are infinitesimal; and if we remember, as we ought to do, that hundreds and thousands of precious lives depend entirely on the *quality* of the Iron used in the Iron rope or chain of our coal-pits, railway engines and trains which cross the country so rapidly, our steam and merchant ships which plough the Ocean and make perilous voyages to all parts of the world—all depend upon the quality of the Iron of which they are constructed, which, if of inferior quality, involves great sacrifice of human life. We could give particulars of pit chains snapping, boilers blowing up, coupling links of railways giving way, the plates of Iron steamers ripping up, crank shafts and piston-rods breaking, all and each involving sacrifice of human life, the recital of which would be heartrending if particularised in these pages. We have paid particular attention to boiler explosions, and can never forget the painful sensations produced on our mind while witnessing the scenes of slaughter and death created over the last twenty years from these causes. The last we saw was Wells of Moxley. We hope we may be spared from witnessing another catastrophe so terrible as this. We believe that if engineers would be more particular in the brands of Iron used these accidents would be much diminished. It is rather a singular fact that we never heard of an ordinary boiler bursting that was made with Lowmoor and Bowling plates, or the plates of a ship ripping up, to the destruction of the vessel and all on board, such plates having been obtained for the said ship entirely from respectable highest-class makers. Such a thing never occurred, and never will occur;—it never could occur, for Bowling, B.B.H., Lowmoor, Earl Dudley's, Snedshill, W. Millington & Co., and S. C. Crown plates might bend with the elementary forces of storms and obstructive rocks at sea, but the hardest rock on which a vessel may be stranded, made of the above plates, could never perforate the hull, or break the plates, or rip the seams. We say this is impossible; hence the importance of our endeavours to impress on the minds of capitalists and engineers how and from whom they may obtain Iron of this class with the view of protecting and saving the lives of the travelling public, who rely on their name and fame for safety. These precautions are paramount because of the difference in the value and quality of Iron. Lowmoor and Bowling to-day are from 30l. to 46l. per ton, according to size and weight, whereas other qualities of the lowest class of Iron, by the lowest class makers, even in Staffordshire, can be bought at 15l. per ton; and as it is impossible to tell what Iron is worth without testing and trying it, the margin in this wide range of value is so great that we believe nothing would be more advantageous to engineers and capitalists than a better knowledge and history of the name, fame, and integrity of the old and highly respectable houses in the Trade, particularly in Staffordshire and Shropshire.   An account of the process adopted at Bowling will be found in this book, and if inexperienced people will ask any experienced Ironmaster, they will be told that Bowling and all other high-class Iron made in Yorkshire and Staffordshire is worth all the money it fetches in the market, much more so than the large quantities of rotten, almost useless cinder Iron made and sold at many pounds per ton less than these brands.   It is a fact that large quantities of Iron are made at certain works from nothing but Cinder Pigs—that is, Pig Iron made from cinders alone, by remelting them in the blast furnace. This is very strange—nevertheless, quite true. Our object in the ' Guide ' has been to point out the best makers of Iron, so that engineers and merchants may be enabled to obtain it through the proper channels and by this means avoid a fearful sacrifice of human life, and a loss of reputation alike to the innocent engineer, and, sometimes, to the merchant who has, perhaps, been deceived.

CONTENTS.

——◦◦——

CHAPTER I.

CHAPTER II.

COMMERCIAL REVIEW OF THE IRON TRADE FOR 1872.

CHAPTER III.

BARROW-IN-FURNESS.

CHAPTER IV.

SOUTH STAFFORDSHIRE OR THE 'BLACK COUNTRY.'

CHAPTER V.

THE RISE AND PROGRESS OF THE STAFFORDSHIRE IRON TRADE.

CHAPTER VI.

CHAPTERS VII. AND VIII

certainly the largest out of London—Chubb's Patent Lock Works—
Edward Davies's Crown Galvanizing Works—Mr. John Morton's
large Warehouse, being the largest Factor in Wolverhampton—The
Iron Exchange—The Town Hall—The Collegiate Church—The
Orphanage, founded by John Lee—The Bilston District Bank,
Wolverhampton—Lloyd's Banking Company, Limited—Punctuality
by Merchants in paying their Accounts at Wolverhampton—Sketch,
with Engraving of Mr. Joseph Wright's Neptune Chain and
Anchor Works at Tipton Green—Cables and Anchors made here for
the Russian Government and the Sultan .

CHAPTER IX.

SHROPSHIRE IRON AND COAL DISTRICT.

Quite one of the oldest centres—Excellent Pig Iron made from
Argilaceous Stone—Mr. John Horton and the Staffordshire worthies,
who first used Shropshire Pig Iron—Coalbrookdale Company—
Fame of this old Foundry—Their Furnaces and Iron Works—The
Kettley Company and Mr. John Williams—The late Mr. William
Botfield at Decker Hill and Malinslee—The Old Park and Sterchley
—Sale of Sterchley—The Snedshill Iron Company—Snedshill Plates
and Charcoal Rods very superior in quality—The Lilleshall Company—Their Nine Blast Furnaces—Their Phœnix Foundry—Engravings of all their Works—Full particulars at page [158]—Shropshire men seldom migrate—The Baldwins of Bilston and Stourport
originally migrated from Shropshire—The towns in this 'Black
Country'—The Madeley Wood Company and the Madeley Court—
Madeley Churchyard the resting-place of the sainted Fletcher—Extraordinary Market at Oakengates on a Saturday night—Wellington
and the Wrekin—Colley and Company's Screws and Boxes made in
Staffordshire   .   .   .   .

CHAPTER X.

NORTH STAFFORDSHIRE.

Robert Heath and Sons, Kinnersleys, were Ironmasters and Bankers
—Under Robert Heath's advice they bought the Clough Hall Estate,
which was found to be full of minerals—Raise the coal at a great profit
—The Bank advances very liberally to the Master Potters—China
trade much encouraged by Kinnersleys' liberality—The late Robert
Heath builds Iron Works at Kidsgrove—The Bank is enriched by Coal
and Iron-making—At the death of his father, the present Mr Heath

CHAPTERS XI. AND XII.

THE MIDDLESBOROUGH, OR CLEVELAND DISTRICT.

CHAPTER XIII.

NORTHAMPTONSHIRE.

CHAPTER XX.

SCOTCH IRON WARRANTS.

CHAPTER XXI.

SPIEGELEISEN, ITS USE AND MANUFACTURE.

CHAPTER XXII.

PIG IRON MANUFACTURED IN 1871.

CHAPTER XXIII.

COAL RAISED IN THE UNITED KINGDOM.

CHAPTER XXIV.

SHIPBUILDING YARDS ON THE BANKS OF THE CLYDE.

CHAPTER XXV.

NEW TRADE OF STEEL CASTING.

CHAPTER XXVI.

WILLENHALL AND ITS LOCKS AND BOLTS.

CHAPTER XXVII.

THOMAS PERRY AND SONS' GREAT ENGINE-SHOP AND FOUNDRY AT BILSTON.

Messrs. George Adams and Son; Messrs. J. P. and W. Baldwin; the
    Albion Sheet Iron Company; Messrs. Lee and Bolton; Messrs.
    William Millington and Company; the District Iron and Steel
    Company; the Snedshill Iron Company; Messrs. W. Barrows and
    Sons; Messrs. William and Thomas Whitwell, Thornaby Works;
    the Darlaston Steel and Iron Company; the Britannia Iron Company;
    Mr. John Marshall; Mr. Henry Hall, Brierley Hill Old Level
    Works; Messrs. N. Hingley and Sons; Messrs. Gold Brothers,
    Sudeley Furnaces; Pontnewynyod Iron Works; Messrs. W. Dawes
    and Sons; the Ashton Vale Iron Company; Messrs. Gyers, Mills
    and Company; Messrs. Appleby and Company; the Chillington
    Iron Company; the Tredegar Iron Company; Messrs. Fletcher,
    Solly and Company; Messrs. Lloyds and Company; the Bilston
    Brook Furnaces; Messrs. M. S. Goddard and Sons; Messrs. Round;
    Messrs. Roberts and Company; Messrs. Watson, Kepling and
    Company; the North of England Iron and Coal Company, Limited;
    Messrs. Thomas Webb and Sons; Messrs. Swan, Coates and Com-
    pany; the Wingeworth Iron Company; Lumphinnans; the Frod-
    ington Works; the Weardale Iron Company; Messrs. Molineux;
    Messrs. Fowler and Company; Messrs. Hugh Martin and Sons; the
    Brymbo Iron Company; the Norton Iron Company, Limited; Mr.
    Robert Crawshay; Mr. John Spencer.
The Earl of Dudley's Works; W. Barrows and Sons' various Works;
    Messrs. John Bagnall and Sons'; the Earl Granville's; the Lilleshall
    Company; the Shelton Bar Iron Company; the Barrow Steel Com-
    pany; Messrs. Bolckow and Vaughan; Messrs. Robert Heath and
    Sons; the Patent Nut and Bolt Company; Cwm Bran Works; the
    Coalbrookdale Company; the Madeley Wood Company and the
    Madeley Court Furnaces; the Bradeley Bridge Charcoal Furnaces;
    Henry Mills; Victoria Works; Isaac Jenks and Sons'; Mayor of
    Wolverhampton's Steel Works; Messrs. Beard and Eberhard,
    Regent Works; Merriman's Lanesfield Works; Messrs. Philip
    Williams and Sons; the Consett Company, and other eminent
    makers, the reader will find specially noticed in the body of the work.

METALS in connexion with the Arts, Civilisation, and Social Progress :—

Gold	Zinc
Silver	Lead
Copper	Quicksilver
Tin	Cobalt
Antimony	Iron
Iron as a Medicine	All the Preparations

Chemicals and Tinctures of the Pharmacopœia made from Iron, adopting the nomenclature of the British Pharmacopœia

A general Table of all the Metals, with Author's remarks thereon, giving chemical equivalents, when and by whom discovered, specific gravity, and their medicinal properties, with commercial use, origin, short history, and general properties

The Table embraces the following Metals ·

Gold	Glucinum	Potassium
Silver	Yttrium	Sodium
Iron	Thorium	Barium
Copper	Magnesium	Strontium
Mercury	Vanadium	Calcium
Lead	Lanthanum	Cadmium
Tin	Bismuth	Lithium
Antimony	Zinc	Silicium
Uranium	Arsenic	Didymium
Titanium	Cobalt	Erbium
Chromium	Platinum	Terbium
Collumbium or	Nickel	Ruthenium
Tantalum	Manganese	Pellopium
Palladium	Tungsten	Niobium
Rhodium	Tellurium	Illemium
Iridium	Molybdenum	Norium
Zirconium	Osmium	
Aluminum	Cerium	

LIST

OF

ENGRAVINGS AND ILLUSTRATIONS.

——✦——

b 2

INDEX TO ADVERTISEMENTS.

ERRATA.

Page 35, line 10, *for* soon after, *read* long before.

Page 106, line 20, *for* the beginning, *read* page 105.

Page 106, line 27, *for* oil, *read* mine and ore.

Page 111, line 12, *after the word* puddling, *add* and.

Page 116, *for* anchors, *read* hoops, *in footnote*.

Page 139, line 25, *for* Sagitarius, *read* Straduarius.

Page 143, *for* Padwick, *read* Paddock, *in footnote*.

Page [163], lines 16 and 17 inserted in error.

Page [173], line 5, *for* Satten *read* Latten.

GRIFFITH'S ANNUAL REVIEW

OF

THE IRON TRADE FOR 1872.

133 CANNON STREET, LONDON.

CHAPTER I.

IN the future annals of the Iron Trade of Great Britain, the year 1872 will always be referred to as the most remarkable on record, not only because of the great expansion of all metallurgical trades and manufactures, but more especially on account of the greatly increased value of Iron which was established in Great Britain in the middle of the year under review. The market price at the time above referred to reached £16 per ton for bars, B.B.H. brand, which was firmly adopted in the fixed rate lists of all the leading Staffordshire manufacturers. This extraordinary revolution in prices will exhibit itself in more striking contrast if we remember that during the year 1870 the official price of bars, £8 per ton, remained unchanged over the whole year: the most important metal produced has therefore increased in value since July, 1871, from £8 to £16 per ton, certainly the most marvellous 'jump' during the space of eighteen months in the value

The price of iron doubled.

B

of Iron or any other metal, to our knowledge, on record.

It is true that in 1806,[1] and especially in 1807, prices of Iron ruled higher than in 1872; but during those years the make was infinitesimal compared with the enormous output of 1872; at this period, Sweden and Russia were the prominent Iron producing countries of the world; now Great Britain produces more Iron than all the world besides, and rules the Iron markets on every Exchange in Europe.

With the above exceptions, never in this century have prices been so high, the exportation and consumption so large, or the general diminution in stocks so marvellously rapid.

The make of Iron for the United Kingdom for 1872, we estimate, after a careful consideration of the whole subject, at 7,250,000 tons. According to Mr. Robert Hunt's valuable statistics, which cannot be too highly commended, the make of 1871 was 6,627,179 tons, which leaves an increase in produce of 522,821 tons over 1872. The value of this 7,250,000 tons, at the present

market price, is about £50,000,000 sterling, taking the mean average price of all makes at £7 5s. per ton, which will be under the mark, if the higher values of the contributions to the great aggregate, furnished by Whitehaven in Cumberland, Barrow-in-Furness, and the valuable produce of the West Riding of Yorkshire, Shropshire, and Staffordshire, are taken into consideration. In producing [2] 7,250,000 tons of pig Iron, at

[1] See Appendix, Table, page 181.
[2] The conclusions we are aware are open to comment, and some good

least [1] 25,000,000 tons of coal, dross, and slack have been consumed, and in foundry, steam engine boilers, puddling, mill-furnace, Bessemer converting, and other processes incidental to the manufacture of Iron, before it is finished into a bar, hoop, sheet, or wire rod, a further consumption of at least 20,000,000 tons must have been absorbed during the year under review. The output of coal for last year we estimate at 125,000,000 of tons, 1871 was 117,352,028 tons. Besides the above large amount of coal consumed in Iron making, the consumption in Iron manufactures, say, nuts, bolts, wrought nails, frying pans, firearms, and other Birmingham wares; the Sheffield, Wolverhampton, Dudley, West Bromwich, and Darlaston hardware goods, with the great edge tool works, engineering and foundry establishments throughout Great Britain, we estimate have consumed another 15,000,000 tons of coal. If we say 25,000,000 for pig Iron, 20,000,000 for the manufacture of malleable Iron and Steel, and 15,000,000 of tons for iron manufactures of all kinds, this gives a total absorption of coal in the Iron trade, for 1872, of no less than 60,000,000 [2] of tons, which, taken from the

Total output of coal

Gross quantity of Coal used in Iron and trades relating thereto.

authorities may hesitate to sanction our rate of consumption, we have the greatest respect for Mr. Bell's opinion, no doubt the highest living authority, and certainly the most practical. The indisputable facts contained in the returns furnished to us by the Scotch, Shropshire and Welsh Ironmasters bring us under the convictions represented in the above figures.

[1] We are aware that we make the consumption much greater than some other greatly esteemed very high authorities, but think careful investigation on the subject will sustain our figures.

[2] See the quantity of coal raised in every county in Appendix tables, page 182.

output above referred to, is about half the whole quantity raised out of the crust of the earth. If we bear in mind the scarcity of coal on the Continent, and the consequent greatly increased demand upon us for this mineral from abroad, the above startling figures will explain the increased demand for this article in our domestic market, and the enormously advanced price established for coal in all parts of the world during 1872. Much has been said and written on this subject respecting combinations and conspiracies among the coalowners, to raise and keep up the price : these figures, however, may perhaps enable even the most inexperienced in these matters to perceive, that the present price of coal has been brought about by increased demand, which must be placed more to the account of the augmented consumption of this article in Iron making, than any other absorbing element either at home or abroad. We have in the United Kingdom, 916 blast furnaces erected, the principal districts being Cleveland, including North, East, and West of England, 290 furnaces, South Wales, 188 ; Staffordshire, 207 ; Scotland, 154 ; total in all districts, 916. Of these, 696 were in blast, or, in other words, at work producing Iron.

In 1861, the number of furnaces in blast were :[1] 565.

We arrive at the above conclusions with regard to the produce of 1872 by careful observation and analysis of returns of the various districts. First, Scotland shows a falling off, to a considerable amount, in yield ; this, however, will prove almost a solitary case.

[margin note: Price of Coal raised by the demand in the Iron trade.]

[1] See " Griffith's Statistics " for that year.

Notwithstanding the difficulties experienced in South Staffordshire from want of coal, the works, with the exception of about five or six weeks, were actively employed all the year, consequently no diminution in the produce is expected. The same may be said with regard to Shropshire and South Wales. We know that an increase of 84,733 tons has taken place in the Cleveland district, and if we give credit in the general estimate for the great activity in the Whitehaven, West Cumberland, Barrow-in-Furness, and other districts in Lancashire, Yorkshire, Derbyshire, and not forgetting the progressive condition of North Staffordshire and Northamptonshire, we think our friends in the trade will agree with us, that the estimate we have made cannot be far wrong.

We have exported Iron and Steel of all kinds last year, 3,388,622 tons; we exported in 1871, 3,169,219 tons, which gives seven per cent. in quantity in favour of last year. The value of the later year's export was £36,060,547, the value of exports for 1871 was £26,124,134, so that the value of the Iron exported from the United Kingdom in 1872, shows an increase of thirty-five per cent. over that of 1871.

If we for a moment consider the innumerably great trades and interests in which Iron forms the chief staple, such as steam engines, all kinds of machinery, railway appliances, the cutlery of Sheffield, the locks, trays, and cut nails of Birmingham and Wolverhampton; the wrought nails, chains and anchors, nuts and bolts, made in the neighbourhood of Dudley, and thousands of other swarming hives of metallic industries in the Black Country; not forgetting the great Iron shipbuilding yards on the banks of the Clyde, the Tees, the Mersey, and elsewhere, the monster shops and works at Sheffield and other centres in Yorkshire—it requires no stretch of imagination to arrive at a truthful conclusion, that the wages paid to forge men, mill men, furnace men, colliers, miners, and the various handicraft men in Iron, far exceeds the revenue of the country, and in a fiscal and social point of view, contributes more to the wealth, progress, and prosperity of this highly-favoured land than any other industry of which England has reason to be proud.

The figures given below refer exclusively to the quantities and value of pig Iron made in 1872. We must now endeavour to show what quantity of this great aggregate was exported to Foreign Countries, and then deal with the balance, showing to the best of our ability how it was disposed of. In order to arrive at the balance, let us see what the total stock of 1872 really was: we think it will be found about as follows ·—

	Tons.
Make of Iron in 1872	7,250,000
Out of Stock in Scotland	296,000
Out of Public Stocks in Middlesboro'	1,800
Out of Ironmasters' Stocks in Middlesboro'	25,000
Out of Makers and Forge Masters' Stocks in Staffordshire, Shropshire, South Wales, Lancashire, · Yorkshire, Cumberland, Northamptonshire, Gloucestershire, and Durham	300,000
Old Iron and Scraps, Remanufactured, including old Cast Iron Re-melted	100,000
	7,972,800

Of this quantity, which is the positive stock of pig Iron upon which we have been working, 1,332,726 tons have been exported abroad, and the balance has been made into manufactured Iron, or smelted down at the innumerable large and small foundries and engine shops of the United Kingdom. The next question which presents itself is, How has it been disposed of ? The Board of Trade returns show clearly that 3,388,622 tons of Iron and Steel of all kinds have been exported. Of this quantity, 1,332,726 was pig Iron, upwards of 660,000 tons to Germany and Holland, 90,200 tons to France, 193,957 to the United States, and 385,687 tons to other countries. The balance, of the 7,972,800, after deducting the export of pig Iron, will be 6,640,074; the balance, therefore, of pig Iron left in England will be, as above, 6,640,074—a very large amount indeed—as the surplus of our make, after supplying the foreign demand for pig Iron by export.

Exact quantity of Pig Iron exported.

Gross quantity of iron and steel exported.

Our next object will be to show how this large balance was disposed of, and mention the various home manufactures and trades which imperatively require it, having, in fact, absorbed and literally melted it away at a considerable profit to the nation. The great home consuming element is of a fourfold character :—

I. Malleable Iron and Bessemer Steel, the making of which has consumed no less than 4,870,074 tons.

II. Tin plates, being in truth Iron plated with Tin.

III. The foundries in all parts of the United Kingdom probably absorbed 1,770,000 tons.

Gross quantity used up at the works and the foundries.

IV. The thousands of manufactures into which Iron enters consumed the balance of malleable Iron made out of the above large aggregate of pig, *minus* the quantity exported in malleable Iron, Steel, and Tin Plates. The quantity of Tin Plates exported was 2,364,684 cwts.

The mills and forges with their 6,841 puddling furnaces and the fifteen Bessemer Steel works with their seventy-eight converters, have, without doubt, made the largest demand upon the stocks, as will be seen by the figures above.

The malleable Iron made at the rolling mills, including the quantity exported before referred to[1], has all been exported or used up for home consumption.

Black plate for tin plates. A very large portion has been consumed in rolling the black plate for Tin Plates. We shipped in 1872 2,364,684 cwts. of Tin Plate, and no doubt consumed in this country three times this quantity. It is impossible to say with precision what quantity of our stock of 1872 was melted by the Iron founders.

In Scotland, however, it is correctly ascertained that the Iron founders consumed 270,000 tons of pig Iron. What quantity has been consumed by the Iron founders in England and Wales we cannot correctly state; we can affirm it to be very large, perhaps, 1,500,000 tons, which, with Scotland, would absorb in all for foundry purposes 1,770,000 tons, which being added to the 1,322,726 tons exported, gives a total of 3,102,726 tons for export and home foundries, leaves a balance

[1] 1,845,351.

for manufacture, out of the gross stock which has been consumed in the Bessemer Pots and puddling furnaces, of 4,870,074 tons.

In the fourth place we come to the most interesting part of the home consumption question, for although element No. 1 consumes by far the largest portion of the pigs, element No. 4 re-consumes and works up the malleable Iron and Tin Plates produced, and fashioning them into implements and weapons of war, articles for culinary and other domestic purposes, and a thousand other tools and articles for the use of man in all parts of the world. Indeed, it is after all, our great manufactures in Iron, Steel, and hardware goods which Shipped North pre-eminently in a commercial point of view places this America country above all others. A plating bar, B.B.H.,[1] which and the Brazils. costs £18 per ton, is made into a beautiful bright axe and sold from £65 to £70 per ton, and the same principle applies sometimes in a greater, often in a lesser, degree to all articles manufactured from malleable Iron, which is the base of Steel; all the knives, saws, and other carpenters' tools of Sheffield, and the cut nails, japannery, and steel pens of Birmingham, the locks of Willenhall, and the saddlers' ironmongery of Walsall, the forged nails of Sedgley, Gornal, Dudley, and the Lie Waste, and the massive cables and anchors of Tipton and Westbromwich, are produced from this article; but there are other large consuming works, among which must be mentioned the great galvanizers of Wolverhampton and Birmingham. This has become

---

[1] B.B.H. is the brand of W. Barrows & Sons of the Bloomfield Works, the best makers of plating iron in England.

a very large trade, which consumes more sheet Iron than any other metallurgic industry in England.

The cut nail makers of Birmingham and Wolverhampton are likewise large consumers of sheet Iron. Besides all these consumptive elements we have our home railways, which are always wanting and continually buying rails, bars, and use Iron, with castings for turntables and signal stands, rails and chairs; the best malleable Iron for their fitting shops, Iron for their bridges, Iron for their stations, Iron for their engines, their wheels and axles, in fact Iron for everything. A very large demand is likewise made upon the wire mills of Shropshire and elsewhere for wire rods for the telegraphs and agricultural fencing in all parts of England and the Continent; and lastly, the Iron ship building yards make a marvellous demand upon our production. The quantity used now, although only a modern outlet, is beyond conception, and as this trade is increasing rapidly, the demand may fairly be expected to continue and increase considerably. Nothing can be more interesting than to see the activity at the gigantic yards, on both banks of the Clyde and elsewhere in this growing and most important Iron industry of the British Empire. Iron is not only in our steam ships the pioneer to commerce, but in every social phase is the willing and useful handmaid of civilization at home and abroad. Our floating docks counteract the disturbing element of the tide in the mighty deep. Iron ships defy the dangers and steam triumphantly over the mountain waves of the Atlantic, gallantly making their way into foreign harbours, to be greeted by thousands of spectators sustained, perhaps, on Iron floating landing stages. We make Iron houses, Iron chapels, Iron shop shutters, Iron girders for houses, and now Iron roads for the trams, which are fast superseding the old inconvenient omnibuses. If we perforate the earth in search of its mineral treasures, the first sod is turned with Iron, the pit is sunken with Iron implements, holes drilled in the hard rock for blasting powder or dynamite to remove the obstructions are made with Steel; roads in the lower depths of the coal and Iron levels are made of Iron ; the waggons of Iron, the cage is made of Iron, the rope is made of Iron, the wheels and the steam engine are all made of Iron, and the brilliant gas which now illumines my pen could not come into the office without Iron. " *Ab initio* " the retort is made of Iron, the purifier of Iron, the gasometer is made of Iron, the great mains of Iron, and the small gas pipes of wrought Iron. It is true the coal yields the gas, but this wonderful production, which seems to defy and subjugate the very laws of nature itself, and by its illuminating power turns night into day, could never have been perfected without Iron appliances from beginning to end. We have endeavoured to cite some of the purposes for which Iron is used, and which have absorbed the 4,870,074 tons in the home trade of pig Iron consumed in the puddling furnaces and Bessemer Pots and made by them into malleable Iron and Steel during the last twelve months. There are hundreds of purposes for which Iron is used, which cannot be mentioned in a work of this kind.

Suffice it to say, we live in an Iron Age, and we believe that Iron will continue to be introduced into new constructions and manufactures, at present unpremeditated, either by the engineer, the architect or the artizan workers in Iron, in this or any other countries.

To assist the reader we recapitulate in the following synopsis the figures embodied in the above statistics :—

		Tons
Gross quantity of Iron made in 1872		7,250,000
Add to which, Stock in Scotland absorbed during same year	296,000	
Out of Store in Middlesboro'	1,800	
Out of Ironmasters' Stocks in Middlesboro'	25,000	
Out of Makers of Forge Masters and Iron Founders' Stocks in Staffordshire, Shropshire, South Wales, Lancashire, Yorkshire, Cumberland, Northamptonshire, Gloucestershire, and Durham	300,000	
Old Iron and Scraps Remanufactured, including old Cast Iron Remelted .	100,000	
Therefore, the gross Stock was in 1872		7,972,800
Exported of this Pig Iron . .	1,332,726	
Consumed in Foundries 'in the United Kingdom .	1,770,000	
Balance consumed by the Puddling Furnaces and Bessemer's Converters and made into Malleable Iron	4,870,074	

On reviewing the Iron Trade for 1872, the cause of the extraordinary rise in price must not be looked for exclusively in the commercial incidents of that year. Certainly without the great volume of demand which came upon us in 1872, £16 per ton could not have been reached ; at the same time it will be well to remember that from 1866 to 1870, was perhaps as flat a period for the Iron Trade as we can recollect, prices during this interval ruled very low, placing Iron at the disposal of architects for constructive purposes, at available prices. During this period girders, bridges, and Iron roofing began to be adopted more generally. The long run of low prices (invariably the case) gave the merchants in all parts of the world an impression that prices would go lower, and under this illusion one and all willingly reduced their stocks to the lowest ebb. In the beginning of 1870, buyers were more disposed to operate, but the subsequent advent of the Franco-German war, cast a gloom over all great staple trades, merchants and consumers of Iron became more cautious, adopting the old tactics of adhering to the reduced stock system, adopted over the previous four years. During the year of the war trade remained quiet, just holding its own without fluctuation or change in the prices of bar Iron during the whole of the year 1870. Although it was not perceived at the time (the course of the trade since has proved that), at that period the consumption of Iron was quite equal to the supply the steadiness of the market during the whole year may be taken as a proof of this fact. In the beginning of 1871, although the war was raging before the walls of Paris, it was thought that better prices might be obtained as the year progressed: puddlers were now only receiving 8s. 6d. per ton in Staffordshire. The Gentlemen of the Streets completed the term of their directorate of France at Bordeaux, peace was made by M. Thiers, and all European nations began to think more of applying their capital and energies to the development of trade and commerce, and were willing to forget the horrors of that devastating conflict by which our gallant Gallic neighbours suffered so .much in men and money. A large trade was done during 1871, with advancing prices in the closing months of the year; and although the incipient stages of a favourable reaction were apparent on all sides, old merchants had no idea of the expansion and increase of value in store for 1872. The rail makers ın Wales, and the pig makers in Middlebro' at this time sold freely both rails and pigs, probably under the influence of timidity in regard to the Money Market, which the enormous French loan, then on the market, looked likely to derange. The Alabama question, prominently revealed just then, likewise a dark cloud in the political horizon, which no doubt contributed to the large sales of pig and rails above referred to. The general activity which prevailed in all departments at the beginning of 1872 imparted considerable stimulus to mining enterprizes in all parts of England, and efforts were made to raise a supply of Iron ore adequate to the smelting powers and requirements of the 916 furnaces capable of receiving the blast. By the increased energy of mine owners, the output [1] of 1872 has reached 19,000,000 tons of ore, against 16 000,000 for 1871, and much larger quantities of Spanish and Elba ore have been imported than heretofore for the use ot the South coast, Welsh and Scotch smelters. The efforts of the masters, however, we regret to say, have

---

[1] We estimate this increase.

been to some extent paralyzed by the unwillingness of the miners to co-operate manfully, in furnishing the required supplies ; it is fair, however, to mention that this dogged indifference to regular and constant working, applied more to the men on the West coast and in Scotland, than other districts ; the miners in Northamptonshire, East Lancashire, and West Cumberland, notwithstanding that labour in the two latter districts is perhaps more scarce than some others, have worked more regularly, received good pay with thankfulness, and performed fair work for their money without grumbling.   The hæmatite ores of these districts having been turned out in much larger quantities, materially assists in swelling the aggregate output to 19,000,000 of tons.   This circumstance is the more gratifying from the fact, that the hæmatite of North Lancashire and Cumberland is decidedly the most valuable raised in the. kingdom, the price ranging last year for good qualities has been over 30s. per ton, the demand continued unabated up to the close of the year.   The famous Park Mine, the property of the Great Steel Company at Barrow, brought to grass no less than 365,000 tons during the year, the same Company, by this and other mines of theirs, turned out in all 660,000 tons in the year.   Much might be said in favour of other splendid properties, among which may be mentioned the Salter and Escot Park, The Millom, The Escott Park, The Park, and various others in the Hæmatite " Eldorado " of Frizzingdon, where the best ore is raised, and the best Iron in the world made, for the manufacture of a peculiar Iron and Steel properly

Probably the most valuable iron mine in the world.

Park mine raised 365,000 tons.

Most district in Cumberland.

named after our illustrious neighbour, Mr. Henry
Bessemer, of Denmark Hill, who invented the wonderful process which has already dispensed with puddling in its manufacture, and produces an article both
of Iron and Steel of superior quality ; and is now being
rolled into rails, and tyres, which readily sell at fabulous
prices, and must in the end cause the name of Bessemer
to be cherished by metallurgists of future ages with
honour and respect.   For while the Bessemer process
acknowledges and accepts with gratitude the efforts
of Dud Dudley, and the late Abraham Darby's
introduction of coal for smelting ; Cort's[1] invention
of Groove Rolls, and Neilson's hot blast ; Bessemer
discovers fuel in the Iron itself, which with inexpensive oxygen gas, increases the heat without cost of
coal or coke, and by his patent process burns out of the
metal all impurities, leaving in the cauldron pure Iron
or Steel charged afterwards with any given *quantum* of
carbon necessary to regulate the quality of the latter.
Notwithstanding the great success which has attended
this process ; we believe it is at present only in its infancy ; in another few years, instead of seventy-six boiling cauldrons vomiting forth in England their spangled
stars of molten silicon and metal ; if we live, we shall
have to report these melting fiery boilers by hundreds,
perhaps in after years by thousands ; so admirably is
this system adapted to develope the Iron trade of this
country.   Before we leave this part of the subject, we
may mention that we have now at work seventeen

---

[1] Cort invented the puddling furnace, by no means a good one, and
but for a subsequent invention of the Refinery, would have been comparatively useless.

Bessemer works in this kingdom.[1] One of these concerns alone, the Great Barrow Steel Company, converted 130,000 tons of pig Iron into Steel, out of which the same works rolled during the year 104,000 tons of Steel rails, tyres, &c., and manufactured at their own sixteen blast furnaces all the pig Iron consumed in this enormous produce of rails, &c., without a single puddling furnace, or puddling in any way. The names of all the Bessemer works, with the converting power of the pots and all particulars of the process, will be found in the Appendix to the " Guide " of this book. Looking at our gross exports of Iron, the shipments of pig Iron preponderate in a marked degree. The Clyde being prominent as a source of supply, the export of bar, angle, and rod Iron, exhibits a decrease in quantity, if compared with 1871; our trade for these kinds with Italy, Turkey, and India, shows a decline on the year, rails likewise exhibit a considerable decrease in quantity. The United States was our best customer, the North American Dominion, Germany, Peru and Australia, coming next. The United States took half the whole quantity of rails exported. The States were likewise our best customers for hoops, sheets, and plates; Australia, India, and the Dominion, consecutively followed the States in quantity for these kinds. Again, the great Republic took two-thirds of all the tin plates we exported, and three-fifths of the large quantity of unwrought Steel shipped by Mr. Isaac Jenks, of the well known Minerva works and others, was taken by the United States of America.

[1] For particulars see Appendix tables.

CHAPTER II.

COMMERCIAL REVIEW OF THE YEAR 1872.

THE price of marked Staffordshire bars in January, 1872, was £11 per ton. The price of Scotch pigs at the opening at Glasgow was 72s. 6d.; in Middlesborough the price was 71s. 6d. At the January quarterly meeting Staffordshire bars were advanced by the leading houses £1 per ton, followed by another advance of the same amount on the 5th of February, which left the price at £13 per ton. On the 11th of April and the 17th of May, consecutive advances of 10s. were made, which left the price at £14 per ton. In June £2 more was added, and accordingly at the next quarter day the price ruled at £16 per ton. A much greater advance in the price of sheet Iron took place than in bars, the *pro rata* scale, which for half a century had regulated the relative prices of sheet with other kinds of Iron, was ignored, and in June and July singles, for prompt delivery, were sold as high as £21, and frequently as much as £23 was paid for Iron of this class; the demand for sheets during these two months being so much beyond the capacity of the Staffordshire mills to roll the requisite supply. In the month of August the demand appeared to falter, and at the close, it was evident to all, that orders were being held back. A general impression now prevailed in buying circles, that the highest point had been touched. A sharp reaction set in; middle men who had bought for the rise, became nervous, and offered their stocks considerably under makers' prices; the action of these speculators alarmed buyers to a great extent, the buying diminished in a marked degree, and on the 1st of October marked bars were reduced £2 per ton; the middle men still undersold the makers at their own doors, which appeared to disorganize the Staffordshire trade in a greater degree, and induced buyers to hold aloof. On the 1st of November a further reduction of £2 was declared, leaving bars at £12 per ton; this fall of £4 had the effect of disorganizing the whole trade. No commensurate reduction having in the meantime been made either in coke, coal, or mine, the pig makers were compelled to reduce the pig Iron to their regular customers, without any reduction in the above minerals, or even wages; and the manufacturers were making and selling Iron at £12 per ton, without any abatement in either coal or wages, which presented in many cases a clear loss on the working of the mills and forges. Most of the manufacturers reduced their make considerably, and put the forges and mills on short time for want of orders. Notwithstanding the reduction, except at a few of the leading houses orders were scarce; thus matters progressed until the end of November, the second class makers working at an absolute loss. As December opened the impression gained ground rapidly among the merchants that prices had touched the lowest point. Numerous orders were simulta-

The highest price reached in July.

Iron had now fallen in value quite £5 per ton.

neously thrown on the market; the general feeling changed for the better, the market hardened, the makers became indifferent to large orders, and although no official advance was declared, prices rallied all round.   Second class Iron went up during the month

from £10, the lowest point, to £11 5s.   Marked bars, although really worth more money, remained at £12, without any official change except notices from the Earl Dudley, W. Barrows & Sons, and John Bagnall & Sons, that further orders could not be taken, except at prices ruling when such orders were executed.   During the same month some misunderstanding appeared to exist with the puddlers and millmen, and the masters, in South Staffordshire; this we are happy to say has now been amicably settled by the masters and men themselves.

In the Middlesborough district the makers have been hampered to a great extent by the miners over the whole year, the furnaces having often been on the point of damping down through scarcity of Iron ore, which the miners persistently refused to raise, except on the hand-to-mouth principle.   Coal and coke have more than doubled in price, railway trucks scarce, and difficulty experienced in the locomotion of metal from furnace to port, shipments were often delayed through the inability of the railway company to serve the increasing necessities of the district.   The Iron-workers have been kept tolerably steadily at work, all difficulty on the wages question having been settled here from time to time by arbitration.

There are 130 furnaces in blast in the Cleveland district, and 19 new ones in course of erection:

1. The Lackenby Iron Co. are building one new furnace. 2. Bolckow, Vaughan & Co. are building one new furnace at Eston. 3. Cochrane & Co. are building one new furnace. 4. W. Whitwell & Co. are building two new furnaces. 5. The Consett Iron Co. are building one new furnace. 6. Rosedale and Ferry Hill Iron Co. are building two new furnaces; Downey & Co. are building two new furnaces at Coatham Iron Works; the Tees Bridge Iron Co. are building two new furnaces; Robson, Maynard & Co. are building two new furnaces at the Redcar Iron Works, Coatham; T. Richardson & Co. are building three new furnaces at West Hartlepool. 7. Hopkins, Gilkes & Co., Limited, are building two new furnaces; most of these being still larger than those lately introduced into this district. The make of pig Iron for the year is 1,968,972 tons, against 1,884,239 tons made in 1871; the reduction in makers' stocks is 25,000 tons, out of public stock 2,800: this make exhibits an increase of the year 1872 over 1871 of 84,733 tons. Prices ranged during the year between 71s. 6d. and 122s. 6d. per ton; most of the makers, however, had sold fair quantities for forward delivery at earlier prices, and on this account were unable to avail themselves to a great extent of sales at the highest figure; considerable sales were made for deliveries of Iron by the makers for 1873, at prices ranging from 82s. to 100s. for No. 3. No. 1 has been scarce all the year, owing to the German demand. At the end of December No. 1 was marked 108s. to 110s. F.O.B. For manufactured Iron a steady trade has been done over the whole year; the

Greatly increased make in the Middlesborough district.

makers in Cleveland were comparatively free from excitement during the months of May and June, and booked orders at from £10 to £12 10s. per ton for bars; plates from £11 to £12 10s., which kept the works going over the months of October and November, while the Staffordshire houses, at the same time, were suffering from a sharp reaction of prices in the general market. South Wales and Monmouthshire progressed steadily in the manufacture of bars and rails, without interruption, throughout the year 1872, the works being fully employed on rail bars, which are the great staple of the South Wales district. Owing to the great proportions of some of these establishments, large orders were booked by most of them in the early part of the year, at from £8 to £9 per ton. Coal, in the meantime, had advanced 50 per cent., which has very much curtailed the profits of these establishments; indeed, the year's work, to several rail-making firms, has been anything but satisfactory in point of profit, and as was justly remarked by one of the greatest makers, the profits this year have been insignificant. Several considerable advances in wages having been conceded to the colliers and Iron workers in this district during the year, and as the demand for rails fell off in the later months, the Welsh masters gave notice to reduce wages ten per cent.; these notices expired at the end of the year, when the men struck for the old prices, and have remained on strike since that time.[1] The price of rail bars ranged over the year from £9 to £11 10s. per ton; some few sales were made at £12 10s.; rails began at £8 5s., at which

[1] We are writing in January 1873.

figure or a little more, orders for the bulk of the make were taken in the early part of the year. For June and July the price was nominally called £11 to £12 10s., but very few, if any, orders were booked at these figures by the large houses. The works are at this moment [1] very nearly all standing for want of coal, through the colliers' strike, which appears, as far as we can judge, likely to continue. There have been much larger importations of foreign ore to South Wales during this than in any former year, the new imports being principally from Spain. The export of rails fell off considerably, 947,548 tons only were exported in 1872, while 1871 absorbed for foreign countries 981,197 tons, giving an excess to the export of rails of 1871 over 1872 of 33,649 tons; this falling off in the foreign demand explains the scarcity of orders, after the highest prices were declared. The make of pig Iron in Scotland in 1872 was 1,090,000 tons, which is 70,000 tons below the produce of 1871, 106,000 tons less than the yield of 1870.

This strike collapsed on the 18th of March, men submitting to masters terms.

Falling off in the export demand for Rails last year.

Serious decrease in the quantity made by Scotland

The deficiency in make of 70,000 tons, with the average price for the year at 101s. 10d. per ton, or 43s. above the mean price of the last twenty-seven years, is a fact worthy of notice, and one furnishing matter for serious consideration. High prices invariably stimulate production, and *vice versâ*. Here, the highest prices accompany a large reduction in the make during a year which has witnessed unparalleled demand, both for home and foreign consumption. The cause of this decline in the producing powers of Scotland, up to this time our great emporium for pig Iron, is a subject of

[1] January 1873.

sufficient importance for the attention of economists and statesmen. It is a mistake to suppose that there is any scarcity of Ironstone in Scotland ; this is not so, and therefore cannot be the cause. Ironmasters have abundance of mineral in the crust of the earth. The canker worm, which seems to be undermining the progress of the Iron trade in Scotland, is the dilatoriness and unreasonableness of colliers and miners, who have kept the Scotch masters in fear and uncertainty over two-thirds of the last year. If these men continue their present vexatious course, Scotland cannot expect to hold her own against other districts here, or in foreign countries, where men work regularly, and when they are well paid, conform to reasonable and honest rules for the good of themselves and their masters. The number of furnaces in blast during the year was 127, the same as 1871. The number of furnaces in blast on the 25th of December was only 115 ; 12 having been damped down in consequence of the strike of the colliers, which reduces the make since the damping of the furnaces at the rate of 78,000 tons per annum. The stocks in Messrs. Connal's [1] great store, on the 29th of December, amounted to 106,919 tons; makers' stocks, 87,081 tons; total stock left in Scotland, 194,000 tons. In 1871, the same date, Connal's stores held 359,860 tons; Forth and Clyde Canal Company's stores, 12,865 tons; in makers' hands, 117,275 ; total stock in 1871, 490,000 tons ; which shows an absorption out

Cause of the falling off in the make.

---

[1] To-day, August 8, 1873, the stocks at Connals are only 44,800 tons.

of stock of 296,000 tons during 1872, leaving the enormous deficiency in the same year of the stocks of pig Iron in Scotland, at no less than 296,000 tons. The available balance now in Ironmasters' hands and Connal stores, only 194,000 tons. This circumstance, coupled with the increased exports of 1872, and the decline in the make of 96,000 tons, together with the prospects of good demand for 1873, furnishes sound data from which conclusions may be drawn, with regard to the average price of Scotch pig Iron for 1873. The tables below, copied from the synopsis of Messrs. Feldtman, of Glasgow, which may be implicitly relied on, will perhaps render the large figures above more easily understood :—

Official Statistics of Scotch Pig Iron.

		Tons.
Stocks end of 1871 .		490,000
Production during 1872 .	. . .	1,090,000
		1,580,000
Consumption in Foundries, 1872	270,000	
do. Malleable Works ,,	200,000	
Exports, Foreign ,,	616,933	
do. Coastwise	224,695	
do. by rail to England ,,	74,372	1,386,000
Stocks end of 1872	.	194,000
1872, decrease		296,000
Stocks end of 1870		655,000
Production during 1871		1,160,000
Consumption in Foundries, 1871	275,000	
do. Malleable Works ,,	190,000	
Exports, Foreign ,,	512,479	
do. Coastwise ..	303,494	
do. by rail to England ,,	54,027	1,335,000
Stocks end of 1871 .	. . .	490,000
1871, decrease .	. . .	175,000
End of 1869 .	. . . .	620,000
During 1870 .	. . . .	1,206,000
		1,820,000

Official Statistics of Scotch Pig Iron (continued).

		Tons.
1870,	298,000 tons	
	208,000 ,,	
	388,842 ,,	
	230,984 ,,	
	35,174 ,,	1,161,000
Stocks end of 1870		665,000
1870, increase .		45,000

Feldtman & Co., Iron Brokers, Glasgow.

During the year under review, the foundries in Scotland have taken 270,000 tons, and the malleable Iron works 200,000, for consumption; exported abroad, 616,955 tons; ditto coastwise, 224,695; ditto by rail to England, 74,372 tons; which gives 1,386,000 tons as the total clearance of Iron sold and delivered during 1872 in Scotland. The table, page 28, will show the countries abroad which have taken the largest quantities of Scotch iron.

We have had great fluctuations in prices on the Glasgow exchange, during the year under review, the lowest (in February) was 72s., the highest price, established in July, reaching 137s. 6d., on the 25th of that month. A reaction now commenced, which continued, with one remarkable interruption, of short duration, until November, when the price descended to 87s. 6d. In this month, a steady advance commenced, which, by the 31st of December, had reached 119s. 6d. cash, closing with higher prices in prospect. The great change in the general market value of Iron, during the first six months of the year, accounts for the advance in this commodity, and the marvellous oscillations in prices which we have witnessed during the year are due in a great measure to the idiosyncrasy of those, who had difficulty in discarding old traditions on the one hand, and spirited buyers on the other, who saw looming in the distance results which must follow the rising value of Iron on the market, in the face of the reduced stock, increased consumption, and a craving demand for Scotch Iron for export, to an extent never before witnessed. The conduct of the great operators for a rise has been criticized, *sometimes perhaps unjustly.* The market was open to all, and the results now prove that the high prices paid were only what the Iron was worth, and is worth to-day, and still continues to fetch these high values readily in the forges and foundries of this and other countries. It must not be forgotten that no make of pig Iron in the world can replace Scotch, for foundry purposes. Everywhere, as a mixture, for reasons known to all, it is indispensable; a little Scotch No. 1 will give fluidity in the casting pot to inferior lower numbers of conti neutal makers; hence the constant demand for Gart sherrie, Coltness, Langloan, Summerlee, or Shotts No. 1, at almost every foundry of note in Germany, France, Holland, our colonies, and the Great Republic of America. This demand must always exist, in the same ratio, until some other district can produce Iron, equally valuable for foundry purposes, to the splendid brands above referred to. During the month of January, in which we write, the Scotch market is well sustained, prices having reached 138*s.* 6*d.* for warrants, cash. The tables below will enable the reader to see at once the different countries which have been the best customers for Scotch Iron, the large figures, which re-

Scotch the best melting Pig for general foundry work in the world.

present Germany, United States, and Holland, being
principally made up of Scotch.

Board of Trade Returns, 1872, *showing the Export of Pig Iron
of the United Kingdom.*

Articles, and to what Countries Exported	Month		Year	
	1871	1872	1871	1872
PIG IRON	Tons	Tons	Tons	Tons
To Germany	9,635	19,917	203,284	313,477
,, Holland	13,188	40,659	246,092	349,405
,, France	5,787	7,526	71,265	90,200
,, United States . . .	8,953	8,111	190,183	193,957
,, Other Countries . . .	18,398	25,022	346,634	385,687
Total . .	55,961	101,235	1,057,458	1,332,726

The trade in North Staffordshire has progressed
steadily, the mining department having had the most
prosperous year on record, and made good profits.
The universal complaint of a scarcity of coal has often
been heard here, more, however, to the inconvenience
of the china and porcelain manufacturers[1] than the Ironmasters. The masters here were more fortunate than
their neighbours in the South, in avoiding large orders
at the lowest prices. We know of one large firm
here who also sold freely at high prices in July. The
wages question has likewise created less difficulty for
the masters, than in Scotland and Wales. A large
business has been done during the year, in hoops,
plates, and bars, but on the whole remunerative prices

---

[1] We can state as a fact that one manufacturer, so urgent was his
necessity for coal to complete the firing of one of his ovens, that he
seized a cartload in the street, and bought horse, cart, and coal from
the owner in order to have the coal.

have been realized, but it must not be forgotten, that even in a highly favoured district like North Staffordshire, greatly advanced prices in the article produced do not yield profits to the manufacturer in the same enhanced ratio.

The output of ore for Northamptonshire, during 1872, will be 1,000,000 tons, by far the largest annual yield ever known. With the exception of one concern, which is idle altogether, the furnaces have worked regularly, and the output of Iron will considerably exceed that of 1871. The demand over the year has cleared off the makers' stocks on the pig banks of the furnaces, which are left now lower than ever they were. The greatest activity has prevailed during the whole year, in the famous hæmatite districts of Lancashire and West Cumberland, and a much larger output of hematite ore has been brought to grass than on any former year. One concern, the great Barrow Steel Co., raised from the Park Mine at Barrow alone 365,000 tons of ore; the mines at Millom and the Frizzingdon District have been more productive this year than ever, and as the quality is improving, rather than otherwise, the ore is eagerly sought after by the Iron trade in all districts. All the pig makers have been particularly busy, and fabulous prices have been paid all the year for the famous hematite makes which are available for the market; particularly those produced in the Barrow and Whitehaven Districts, these pigs being so much in request at the Bessemer works in different parts of the kingdom. The miners have worked fairly, and although labour is scarce, there has been no serious wages difficulty here during 1872. Both for mine owners and Ironmasters, 1872 will be remembered as a prosperous year. Having closed our review of 1872, our friends will expect a word with regard to the prospects in store for the Iron Trade during the year upon which we have now entered. Those who have our " ANNUAL REVIEW OF THE IRON TRADE," for 1871, will, by reference thereto, read the last sentence contained therein, which is as follows :—" From these favourable premises, we augur a year of greater prosperity to the Iron Trade than any on record during the 35 years that it has been our pleasure and privilege to address ourselves in these Annual Reviews to our numerous clients and subscribers." We believe the progress of the year reviewed above has fully borne out our anticipations, which at the time they were written might have appeared a little sanguine. Now, with regard to 1873, the prospects of the Iron Trade are good, and with the exception of the unsatisfactory state of the labour market, in Wales and Scotland, we can observe nothing particularly adverse to the sound progress of the trade in the future horizon, which, by the wisest, can only be dimly seen. It is true, the high price of coal, while the present demand exists, will inconvenience the Ironmasters, and curtail their profits, and no abatement in the price of fuel can be expected, until a much larger output can be brought to grass. We must, however, wait with patience the advent of increased supplies : for the present this is impossible. Therefore, taking things as they are, and judging of the present situation, on all sides, as it is, and appears likely to remain, over the year of 1873, if masters and men work harmoniously together, we believe the Iron Trade of 1873 will be prosperous in a moderate degree, and yield a fair profit to the Ironmasters for their large capital employed therein, the men will have an opportunity of making good wages, and no doubt will make greater progress in moral and social attainments than they have been accustomed to do, under the old scale of pay. The new rate will enable them to perform their social and ministerial duties with good will and alacrity, not only to their employers, but in the hundreds of thousands of small social circles, of which the Colliers and Ironworkers are the mainstay, and chief support. In 1820, the United Kingdom produced 400,000 tons of Iron. In 1826, the output had increased to 600,000 tons, and, as will be seen by the following Table, in '27 it had reached 690,500 tons. The Table will likewise exhibit to the reader the progress made in the interesting interval from 1820 to 1827.

This Table shows the comparative make of pig-iron in 1820 and 1827, taken from the *Encyclopædia Britannica* :

	1820 Tons.	Furnaces.	1827 Tons.
North Wales	150,000	12	24,000
South Wales		90	272,000
Shropshire	180,000	31	78,000
Staffordshire		95	216,000
Yorkshire	50,000	24	43,000
Derbyshire		14	20,500
Scotland	20,000	18	36,000
	400,000	284	689,500

Mr. Kenyon Blackwell in a paper read before the Society of Arts, in 1854, on ' The Iron Industry of

Great Britain,' gives the following figures, which show the make of Iron in all countries for the previous year. [1]

The estimated production of crude Iron in the various countries.

	Tons.			Tons.
Great Britain	3,000,000	Russia		200,000
France . .	750,000	Sweden		150,000
United States	750,000	Various German		
Prussia	300,000	States	.	100,000
Austria	250,000	Other Countries	.	300,000
Belgium . . .	200,000			6,000,000

On referring to the above it will be seen that Great Britain produces as much crude-iron as all other countries together.

[1] The name of Mr. Samuel H. Blackwell and Mr. Kenyon Blackwell, of Dudley, are well known as the best authorities on this matter.

cn show
'.. year. 1
1 in the

T...
2.0,00
150,000

1.0,00
50,00
...0,00

at Great
ler coun-

n Black-
.s matter.

CHAPTER III.

BARROW-IN-FURNESS.

WE are quite in the dark in regard to the exact period of time when the rich hematite ores of Barrow-in-Furness were first smelted into Iron. There is however good reason for believing that the ore of this district was worked in the neighbourhood of Barrow-in-Furness anterior to the manufacture of Iron in Sussex, which is well known to have been the principal Iron county of England during the ages prior to the Christian era. This hypothesis receives considerable support from the fact, that the wood in the forests of the Barrow district had been partially exhausted for fuel as long since as the reign of Queen Elizabeth. That Iron making was carried on here in remote times, is beyond all doubt, for the name of Furness Abbey, recorded in Domesday Book, was taken from the Iron "furnesses" in the neigh bourhood; the cinder heaps remain to this day; the remains of the former have likewise been dis-covered and pointed out by numerous enterprising metallurgical antiquarians: we have seen them ourselves. Archimedes of Syracuse, 280 before the advent of Our Lord, destroyed the Roman fleet, under the command of the famous General Marcellas, by immense chains

D

and grappling irons. No doubt Archimedes had greatly improved upon Tubal Cain's process, and he must have been the greatest metallurgist of his day ; the Romans very probably learnt the trade from Archimedes or some of his pupils. Marcellas declared afterwards at Rome 'that by these irons Archimedes treated his galleys as though they were mere buckets to draw water with.' In the early ages of Pàgan Rome, the manufacture of Iron was practised by the Romans, ore was brought from the island of Elba: some of the Roman historians mention this circumstance. All the Iron used by the good Emperor Vespasian in the construction of the Colliseum was made from Elba ore. We have carefully inspected these mines, and reported upon them as being highly valuable and rich in Iron : we landed at port Hercule. Iron must have been very valuable at this period, and during subsequent ages. On a minute inspection,[1] it appeared to us that the partial demolition of this gigantic structure by the early Christians was effected more with the object of disengaging the precious Iron cramps, stays and plates, which hold it together, than for any other matériel originally used in the construction of the building, for this simple reason that Iron was a great necessity and realised a very high price. We observed indications on various parts of the structure of great labour in the surrounding stone-work, to extract the tie-bars which so admirably hold the building together. These efforts were unavailing in those early times, for large masses of Iron, embedded in the

The Ebbw Vale Company import Elba ore to this day.

[1] When in Rome.

stone-work of the building, still remain, having defied the power of the best punches and hammers the Romans could in those days produce. It may be, for all we know, that we owe the existence of this extraordinary memento of Roman grandeur to the strength and resistance of the Iron which holds it together. No doubt Julius Cæsar or Agricola perfected the art of Iron making in this country at the first Roman invasion, for it is known that Cæsar visited the great forges at Syracuse soon after he crossed the Rubicon. Having had opportunities of witnessing the modes adopted in other countries, particularly at Syracuse, Gaul, and Spain, Julius Cæsar and his followers would be able to point out the best mode then known of separating the metal from the matrix. At this time, the ancient bloomeries were introduced in England, a method for making Iron which was followed up to the sixteenth century in all parts of the world; the different shapes and forms being wrought by manual labour, principally with hammers. The blast for the furnace was created by blow-bellows, or a rude piston, fitted into the hollow trunk of a tree, the motive power being manual labour. It is very singular that this process was somewhat analogous in its effect on the iron, and produced in some degree the same results as Mr. Henry Bessemer's process, by a continual injection of blast, for, although the pressure must then have been moderate, it left the finished Iron so far exhausted of carbon, as to be malleable and fashionable into useful forms under the hammer. The next great stride, made by Cort at the end

VIEW OF THE BARROW STEEL WORKS.

of the seventeenth century, was the invention of the puddling furnace, which enabled us rapidly to increase the quantity; but the Iron made during the existence of Cort's sandbottom puddling furnace, was very inferior in quality to that previously produced by the old bloomeries of North Lancashire, Dudley, Sussex, and the Forest of Dean, as no scientific principle was then involved in its manufacture. The utter want of knowledge in this respect left the manufacture subject to great waste in quantity and deterioration of the quality of Iron produced. Cort, however, left a valuable legacy to posterity in his invention of groove rolls. Mr. Joseph Hall, the practical partner at the great Bloomfield Works, invented the present system of puddling, which superseded Cort's plan altogether; and designed likewise, a different puddling furnace adapted to melt and boil the metal in a molten sea of silicon, which protected the Iron from the devouring effects of the oxygen, then constantly playing on the top part of the charge under Cort's process, before the whole mass was sufficiently melted and properly decarbonized. Mr. Hall also perfected the preparation of the present tap cinder now in use, called bull-dog, for making and repairing the inside of the puddling furnaces. These improvements are entirely due to the late Mr. Joseph Hall, of the firm of Barrows & Hall (now W. Barrows & Sons), of the great Bloomfield Iron Works at Tipton, Staffordshire. This brings us to the advent of Mr. Henry Bessemer's improvements for making Iron in 1856, which have given such an impetus to the manufacture of Steel and

Iron, and in a remarkable degree revolutionized the trade. In a few words, the Bessemer process consists in the injection of constant volumes of atmospheric air, at a high pressure, through apertures in the bottom of a great iron cauldron, containing from two to eight tons of the molten metal. This process continues until the oxygen in combustion has devoured, or, chemically speaking, absorbed and dissipated, the carbon ; the silicon being burnt out with the sulphur and other impurities which the Iron contains ; and leaves, as Mr. Bessemer himself informs us, only traces of phosphorus in the iron : the affinity of phosphorus for Iron being so great as to defy the searching fiery ordeal in this truly burning fiery furnace. The hæmatite ore of Barrow and West Cumberland contains scarcely a trace of phosphorus. Our readers will therefore be able from the forgoing to see and understand why Iron made from these ores is and will be so eagerly sought [1] after by all the Bessemer Steel converters in the world; although the price of this ore has more than doubled during the last few years.

The Bessemer process in operation.

A description of the Barrow Steel Works, fixed as they are in the centre of the hæmatite district, will, we think, be interesting, particularly to Ironmasters and others, who like ourselves have watched for years the progress of the Iron trade, this the largest and most important concern in England. His Grace the Duke of Devonshire is chairman of the company ; Sir James Ramsden is the manag-

The Great Barrow Steel Works, with illustration.

---

[1] The Cumberland Hæmatite ore is free from phosphorus. The Barrow Hæmatite ore is likewise without phosphorus.

ing director, and Mr. Josiah T. Smith, the Mayor of Barrow, being general manager. This company was formed about eight years since. The works consist of sixteen blast furnaces (two more in course of erection), eighteen Bessemer melting cauldrons, three rail mills, with hydraulic lifts, one plate mill, one merchant mill, and two tyre mills. These furnaces produced last year 250,000 tons of Pig Iron, 130,000 tons of which they converted into Bessemer Steel and Iron, and sold, made into rails, plates, tyre bars and forgings. The output of this class of manufacture— steel rails, plates, tyres, bars, and forgings—is at the Steel works 2,000 tons per week. They had raised last year 365,000 tons of hæmatite ore, of the best quality, at their Park Mine. The company has another mine, the Stank, two miles from Barrow, which they are now actively proving, and which from all present appearances will rival the celebrated Park's Pocket above referred to.[1] The Barrow Steel Company consumes annually 300,000 tons of coke, and 150,000 tons of coal, in Iron and Steel making, and employs in all departments 10,000 artizans and workmen. They have recently erected most extensive jute mills, the gentlemen of the Steel works being proprietors (Sir James Ramsden is the chairman of this Jute Company), which for extent, perfection of machinery and architectural nobility of construction, rival any mills of the kind in Scotland or elsewhere. At present, 1,200 pair of hands are employed here; this number

[1] At the Stank they had driven out laterally into 16 acres square of good mine, and the drivings are still in ore.

The gigantic Docks at Barrow.

will be increased before the end of the year to 2,000. Two monster docks are already completed, called the Devonshire and the Buccleugh Docks, twenty-four acres each in extent.   Another, the largest however of the three, is now in course of construction, to be called the Ramsden Dock, after Sir James.   It will be completed next year, and will occupy an area of 200 acres, and be quite capable of admitting a vessel 20 feet longer than the Great Eastern. The Duke of Devonshire is chairman of the railway,[1] dock, and Ship Building Company, which is on a gigantic scale. The yard is intended to employ from 6,000 to 8,000 men, and will be capable of turning out one of the largest steamers per month. We understand this company have it in contemplation to open a line of steamers of the highest class, between the ports of Barrow and New York next year, half a million of money having been already subscribed among themselves for this purpose. The market price of the produce per annum of the Iron, Steel, hæmatite mine and manufactures, we understand on reliable authority, taking it at the market value of to-day, would be £5,335,000 per annum.[2] A reference to the above description of these works, for the accuracy of which

---

[1] The railway from Cairnforth to Ulverston, Barrow and Whitehaven belongs to this company, Sir James being the chief manager.

[2] Produce of the Barrow Steel Company and its present value, which is as follows :—

250,000 tons of Pig Iron at 9l. per ton	.	2,250,000l.
104,000 tons of Rails at 20l. per ton, including Forgings, &c. . . .	.	2,080,000l.
600,000 tons of Hæmatite Ore from the Parks, Pocket, and other mines at 33s. 6d.	.	1,005,000l.
Per annum	.	5,335,000l.

we can vouch with confidence, leaves the works of the Demidoffs of Russia, Kruppe of Prussia, Schneider of France, and Cockcrell, Imperial Foundry of Belgium, far behind, and stands out as a striking and continually increasing memento of the supremacy of this country in all metallurgical and Iron industries. Too much credit cannot be given to the noble Duke of Devonshire, for his fostering care and attention to this gigantic company. Next to His Grace, Sir James Ramsden must be highly complimented and praised for his bold conceptions, backed by an iron will to carry them out. Mr. Josiah T. Smith, the general manager, from first to last has likewise, both in the construction and conduct of the works, developed perhaps a more profound knowledge of the business than could be met with in any other quarter. Mr. Smith is the son of a Derbyshire Ironmaster, and for several years was on the most intimate terms with the late Samuel Holden Blackwell, who was admitted to be the most scientific Pig Iron maker in South Staffordshire, where Mr. Smith acquired a thoroughly practical knowledge of Iron manufacture. We were present in 1854, when the first cast of Iron was tapped at the Barrow furnaces, and visited the works again last year. It would be impossible to express the surprise and admiration felt on the last visit, at the truly marvellous progress made. In 1854 we observed only a few straggling houses at Barrow ; now we found a large town, with sixty or seventy thousand inhabitants, oil mills, a well-conducted newspaper, the " Barrow Furness and North Western Times ; "

imposing shops ; hotels ; thousands of well-built houses ; wide streets, running, as they ought, at right angles, the style of the whole being in accordance with modern architectural principles ; elegant churches and chapels of all denominations abounding ; all public works are under supervision of the Corporation, which is ably presided over by Sir James Ramsden, who was chief magistrate for several years.[1] A line of very high-class steamers run between Barrow and Belfast. So great, indeed, was the change, and bewildering to us, that we could not find our way to the furnaces which we had formerly visited. The perfection of the machinery in the Steel works, the ease and regularity with which the rails were turned out, two from one ingot, the finish and quality of the plates rolled, the great precision with which the guardian operator at the cauldron stopped the consuming element when the work was done— all excited our admiration to the highest pitch. The quality of the Steel and Iron for rails, tyres and wheels, being far superior to others, is capable of greater tenacity and flexibility, and susceptible of resisting impact force more than any others. From observations we made, its ductility is unequalled, and calculated to resist in an eminent degree, transverse pressure, tension, and compression ; it has likewise great elasticity and resisting powers to torsion strains, which clearly indicate that it is the most valuable Iron made for rails, wheels, girders and tyre bars.

[1] Since the above was written Josiah T. Smith, Esq., has been elected mayor of Barrow.

Sir William Fairbairn says 'the great advantage to be derived from the Barrow manufacture of Steel is its ductility combined with a tensile breaking strain of from thirty to thirty-two tons per square inch.'

We cannot doubt the correctness of this statement, coming as it does from the highest authority on these matters in England. In our experience, which has been as extensive as most Ironmasters and engineers, we know of no other Iron or Steel which will bear anything like the tensile strain mentioned by Sir W. Fairbairn : the tests we understand were carried out and witnessed by himself. We give a table next page published by Sir William some time since, giving a minute analysis of all the ores which this great company smelt at their Blast Furnaces. The whole quantity having been raised from their own mines.

Analysis of Iron Ores used at the Barrow Hæmatite Iron and Steel Company's Wrks, Barrow-in-Furness, Lancash'e.

No	Name	Water	Sesquioxide of iron	Iron	Phosphoric acid	Phosphorus	Sulphuric acid	Carbonic acid	Silica	Alumina	Protoxide of Manganese	Lime	Magnesia	Insoluble residue	Total	Silica	Alumina	Lime	Magnesia
							Solution									The Residue			
1	Park Ore (average)	1·91	76·77	53·74	0·04	0·02	none	none	0·14	0·04	0·63	0·24	trace	19·79	99·58	18·51	0·99	0·04	trace
2*	„ (best rough)	0·47	94·88	66·42	0·03	0·01	none	none	0·10	0·07	0·04	0·34	trace	4·49	100·39	4·45	trace	trace	trace
3*	„ („ , fine)	0·68	90·44	63·31	none	none	none	none	0·09	0·30	0·30	0·30	trace	9·11	100·95	8·74	0·24	trace	trace
4	Lindal Moor ( )	2·02	78·61	55·03	0·03	0·01	0·04	none	0·04	trace	0·24	0·57	0·19	18·31	100·05	16·11	1·67	0·03	0·05
5	No. 2	1·61	76·05	53·24	0·04	0·02	trace	trace	0·03	none	0·08	0·49	0·14	21·07	99·53	18·60	2·04	0·08	0·11
6	No. 3	2·68	70·17	49·12	0·04	0·02	0·03	trace	0·06	0·37	0·31	0·59	trace	25·24	99·49	22·24	2·48	0·24	0·16
7		10·84	65·21	45·65	0·03	0·01	trace	trace	trace	0·24	1·01	0·41	0·14	22·38	100·26	18·67	3·42	0·05	0·04
8*	Lindal Cote (puddling)	2·82	77·24	54·07	none	none	none	4·19	0·09	0·24	0·11	6·00	0·41	9·07	100·17	7·27	1·47	0·08	trace
9*	Lindal Moor (puddling)— No. 1	3·35	86·20	60·34	trace	trace	0·04	1·43	0·08	0·43	trace	2·23	0·59	6·50	100·85	5·58	0·58	0·05	0·05
10	No. 2	2·35	66·60	46·62	trace	trace	none	5·96	0·13	0·23	0·07	6·64	1·94	16·28	100·20	14·02	1·76	0·10	trace
11*	Whitrigg's (filling)	1·97	83·33	58·33	none	none	trace	2·53	0·04	0·02	0·08	4·05	0·15	7·51	99·68	6·55	0·73	0·05	trace
12	Dalton's (ht.)	1·80	67·14	47·00	none	none	none	4·45	trace	0·25	0·08	6·02	0·15	19·77	99·66	19·09	0·51	0·12	trace
13*	Mouzell Mine (cst.)	2·28	83·94	58·76	0·03	0·01	none	none	0·09	0·22	0·28	0·65	none	13·17	100·66	12·37	0·48	0·20	0·09
14	„ (average)	1·40	69·41	48·59	none	none	none	none	0·06	0·02	0·02	0·44	0·13	27·97	99·48	25·92	1·53	0·07	0·04
15	Newton Mine (blast)	3·08	77·64	54·35	trace	trace	trace	none	0·01	0·15	0·13	1·09	0·14	17·94	100·17	15·44	2·13	0·06	trace
16	Urswick (ht.)	6·59	61·30	42·91	0·02	0·01	trace	none	0·02	0·28	0·24	1·01	0·58	29·73	99·75	26·78	2·54	0·16	0·07

NOTE.—The ... es were not determined.

* Used for making Iron for Bessemer. †.

EARL DUDLEY'S COAL-FIELD.

CHAPTER IV.

SOUTH STAFFORDSHIRE, OR THE 'BLACK COUNTRY.'

IT has been known from ancient times that Staffordshire was rich in Ironstone and Coal. Plott often
refers to this circumstance in his history of Staffordshire. Even in more remote ages, during the Roman
occupation, Iron was manufactured in the neighbour- *The Earl of Dudley's vast mineral domains.*
hood of Dudley by primitive means then in vogue,
charcoal was made from the wood of the dense
forests which at that time overspread the undulating
territory, now in the possession of the Earl of Dudley,
the priceless value of which could not be accurately
estimated at the present day, owing to the inexhaustible
seam of coal varying from ten to fifteen yards in thickness, the best quality in the world for Iron making and
Ironstone, which the crust of the earth contains in this
vast princely domain. The 'Black Country' commences *Topography of the Black Country.*
at Wolverhampton, extends eastward a distance of six-
teen miles to Stourbridge, eight miles to West Bromwich, penetrating the northern district through
Willenhall to Bentley, Walsall, the Birchills, and
Worley; embracing under its darkened canopy of
smoky atmosphere the townships of Wolverhampton
and Willenhall, with their locks and japannery, their *Manufactures of the 'Black Country.'*
curry combs and boiling cauldrons of galvanizing
spelter; Walsall and Darlaston, with their stirrups and

bridle bits, nuts, bolts, and other railway appliances; Wednesbury, with its gas-tubes, foundries, gun-locks, and coach springs; Smethwick and Dudley Port, with a thousand swarming hives of metallurgical industries on the banks of the Rail and Canal Companies, too numerous to mention. In this immediate vicinity we have Chance's monster glass works at Spon Lane; and the great alkali works of the same firm at Oldbury. Here, too, modestly stand the Soho works, so famous in history, where the immortal Watt made his first condensing steam-engine. Here we have likewise Muntz's patent metal works. The great works of the Patent Nut and Bolt Company, the Patent Rivet Company, the Plate Glass Company, and Joshua Horton's boiler yard, of world-wide fame, are all situated at Smethwick.

West Bromwich and Hilltop are contiguous; here enamelled and tinned pots, kettles, and saucepans are manufactured, in all shapes and sizes, on the most extensive scale. Oldbury lives close by, where with bated pulsation, under a constant cloud of black smoke, the vivifying rays of the sun being obscured here by the volumes of almost material carbon floating in the atmosphere. On the one hand there is the destructive effect of the smoke on vegetation, on the other, of the hydrochloric, sulphurous, and chlorine gas evolved from numerous chemical works almost in the heart of this devoted township. Bright grates and fire-irons become rusty in a single night, and all household furniture, which is held together by appliances of Iron, suffers much, and all other metals are damaged by these gases

Marginal notes:

Chance's monster Glass Works. Largest in the World. Bolton & Watts, Soho Engine Shops.

Oldbury in smoke.

Injurious effects of acid gases.

GETTING COAL AT ONE OF EARL DUDLEY'S THICK COAL PITS.

in the same proportion.   Iron mines and collieries sur-
round the town, the workmen, on their return from
work at the pits in the evening, show honourable traces
of the useful labour they have performed, in the soiled
garments and dirty faces they present.   Vegetation suc-
cumbs altogether ; scarcely a shrub, a tree, or a green
field is to be seen, amid the general devastation of the
surface, which presents itself for miles round.   The mo-
notony of the landscape being broken only by irregular
mounds of earth and mountains of furnace cinders, the
former being the disembowelled crust of the earth, removed and brought to grass by the toiling miners, in
search of the valuable Ironstone and ' black diamonds,'
so plentiful in the geological formations of the entire
district surrounding Oldbury.   The heaps of furnace
cinders are the glassy refuse of molten silica, and lime,
which the blast furnace discharges in her process of
separating the metal from its matrix, before the Iron
is consolidated and run out into the pig beds, a considerable quantity of lime being requisite with the ore
and coal in the furnace to facilitate the smelting process.
Nothing to be seen by day but smoke, heaps of furnace
cinders, and abnormal mounds of earth and coals, and
by night, the lurid glare of a thousand burning furnaces
of various colours, from the blood red of the puddling
furnaces, to the yellow and blue flame of the copper
works, and the chequered red and white flames,
emitted in the largest volumes, from the funnel heads
of the Blast Furnaces, which may be seen in the dis
tance all round the town.

Description of Oldbury by night.

 We next come to Dudley, an important town,

Dudley
and its
manufac-
tures.
completely surrounded on three sides by smoke and flame. This is a great emporium for chains, cables, anchors, grates, fenders and fire-irons, and, above all, for wrought nails, which are brought by the nailers from Sedgley, Gornall, Brierly Hill, the Lye Waste and other districts. The nail factors supply the Iron to the men, who produce the nails at a scale of prices mutually agreed upon per pound. Anvils, vices, stove grates, and fenders are likewise made here on a large scale, and a very high class of Grates, Fenders and Fireirons are made at Marsh's works at

Most of
the land
belongs to
the Earl
of Dudley.
Burnt Tree. Most of the land in the neighbourhood belongs to the Earl of Dudley, whose agent (Mr. Fisher Smith) resides at the Priory, situated in a lovely spot, the grounds of which are charming in the extreme. Frequently while walking and talking with the late Mr. Richard Smith, the former agent, in these grounds, as early as six o'clock on a summer morn- ing, we have remarked to him that ' this is truly a

Oasis in
the Black
Country.
lovely oasis in the " Black Country." ' Dudley Castle, a fine old monument of the Feudal Ages, proudly crowns the Castle Hill, in front of the Priory, and is a place of

Dudley
Castle
place of
resort for
holiday
makers.
great resort for tourists and pleasure-seekers, who, by permission of the noble Earl of Dudley, have every facility for exploring this fine old ruin, which stands upon gigantic caverns, often illuminated with gas, on State occasions, to gratify the Black Country people, and, as may be supposed, his Lordship and the noble Countess are very popular with the people. Brockmoor, Brettle Lane, Wordsley, and Stourbridge, must be included in the Black Country group of towns.

UNDERGROUND WORK AT EARL DUDLEY'S SALT WELLS COLLIERY,
Where the Thick Coal is obtained.

Although the atmosphere becomes purer as we get to the higher ground of Brierly Hill (Lord Dudley's famous Round Oak Works are here), nevertheless here also, as far as the eye can reach, on all sides, tall chimneys vomit forth clouds of smoke, and the sulphurous flames of the fiery furnace are observed in all directions. Our feeble efforts to describe these districts will, we hope, satisfactorily explain why it is so emphatically called the 'Black Country.'

In this, the Stourbridge and Brierly Hill district, a very extensive business is carried on in the manufacture of fire-bricks of all kinds, used in the construction and relining of blast furnaces, puddling furnaces, cupellers, and air furnaces. The fire-clay deposits here are reputed the best in England, being fashioned into melting pots and gas-making retorts, which fetch high prices ; these bricks are exported largely, and are highly prized in all parts of the world, particularly the foreign settlements of the British Empire, the general opinion being that they resist the highest temperatures in smelting furnaces of any others which have yet been produced. A list of all the best manufacturers will be found in the Appendix to the Guide. Perhaps Ruffords', Mrs. Emily Gibbons', and Pearson and Harrisons', make the best quality. Mrs. Gibbons, relict of the late Benj. Gibbons, Esq., the well-known Ironmaster of the Milfields furnaces, we are informed by Mr. Jones, of the Commercial Gas Works, here stands unrivalled for the manufacture of these Gas Retorts.

The best retorts for gasworks and fire-bricks in the world made here.

CHAPTER V.

THE RISE AND PROGRESS OF THE STAFFORDSHIRE IRON TRADE.

ALTHOUGH Staffordshire is not the oldest, it has been for three-quarters of a century the most important Iron-making centre in the kingdom. South Wales and Shropshire may, with greater truth, be called the pioneer districts of the Iron trade than South Staffordshire. In 1750 Staffordshire could only boast of 17 [1] blast furnaces, which had increased in 1780 to forty, and in 1806 to forty-two; the exertions, however, of Dud Dudley, whose labours and difficulties are minutely recorded in his little book, were of paramount importance, *ab initio*, in opening the road to Iron industries, since so successfully followed up in Staffordshire and other Iron-making centres; we refer to the introduction of coal at this time for smelting Iron instead of woodash or charcoal; here, however, we may remark that the credit of this appliance must not be awarded entirely to Dud Dudley: Mr. Abraham Darby,[2] of Coalbrook Dale, applied coal in the same way with successful results in Shropshire antecedently to Dud Dudley's ex-

The late Abraham Darby and Dud Dudley

[1] The make of Pig Iron at this period was 30,000 tons per annum; in 1800 it had increased to 180,000 tons; and in '25 the production reached 600,000 tons for the United Kingdom.

[2] We think the gentleman above referred to was the great-great-grandfather of Mr. Abraham Darby, of Coalbrook Dale.

periments, which, we all know, terminated in partial success, notwithstanding the difficulties and persecutions which this original Dudley Ironmaster had to contend with. Nevertheless, Dud Dudley must take the credit of tapping the crop in various places of that most valuable bason of thick coal (from ten to sixteen yards thick in one seam), and applying it to useful purposes in Iron smelting, and which in after years has done so much to enrich the Ironmasters, and facilitate the profitable manufacture of this metal in the Black Country. During the final decades of the last century, Mr. John Wilkinson occupied the most prominent position as an Ironmaster, and had extensive works at Bradley, Hall Fields, and elsewhere, Mr. Reid, Mr. Parker, and the Addenbrookes, at Moorcroft, the descendants of the latter,[1] being still in the trade, played a useful and important part in the early history of Staffordshire Iron making. The names of Samuel and John Fereday, Firmestone, and Foley were likewise well known in connection with it at a more remote period. Mr. James Foster subsequently joined the works at Stourbridge, which were formerly built and carried on by his uncle, Mr. John Bradley, under the style and firm of John Bradley & Co., their principal manufacture at the commencement being nail rods, which were supplied to the districts of Stourbridge, Wordsley, Brierly Hill, Dudley, Sedgely, and Wolverhampton. The nail makers sent their carts to the Works for the Iron ; eventually Mr. Foster arranged to deliver the Iron in horse waggons, and settle the accounts once a quarter ; he was in the

smelt iron with coal successfully for the first time.

The pioneers of the Iron Trade in the Black Country.

[1] Are now working two blast furnaces at Rough Hay.

habit of attending the Lion Hotel, at Wolverhampton, the second Wednesday of the quarter month for this purpose; Mr. Reid, Mr. Philip Williams, Mr. Wheeler, Mr. Addenbrooke, John and Samuel Fereday, Mr. Grazebrook, and Mr. Parker met on the same consecutive days at this place, and by degrees got into the habit of doing their business in the same manner. This was the beginning of the Ironmasters' Quarter

Quarter day first introduced.

Day, which arose and originated through Mr. Foster, about this time. The same name for the settling day has since been adopted by the Ironmasters, Factors, and Merchants in the Black Country, and Merchants from all parts of England continue to attend this great Exchange, which is always held, as above stated, at Wolverhampton the second Wednesday, and Birmingham the second Thursday in the month. In former years the closing meeting was held on Saturday evening, at Dudley, where a good dinner was provided. This, however, has now dwindled to insignificance, the Birmingham Quarter Day, of late years, having become the greatest, and by far the most important, of these now tri-monthly gatherings. Mr. Foster enlarged his power of production rapidly, availing himself of the valuable thick coal which abounded in the neighbourhood, and subsequently purchased the Madeley Court estate in Shropshire, which overlies beds and seams of the best coal and Ironstone in Shropshire, which was smelted then into Pig Iron for the use of the Stourbridge works; this was a great stroke, enabling him to perfect the quality of his iron, which was marked S.C. Crown.. He was in the works continually, and in the

THE BIRMINGHAM EXCHANGE.

The Iron Masters of the Black Country meet here every Thursday. This is the largest and most influential periodical assemblage of Iron Masters held in the United Kingdom. Most of the magnates in the coal trade and great manufacturers meet the Iron Masters here at Half-past Two every Thursday.

early days did not hesitate to put his hand to a bar of
Iron if necessary. He was a most able and far-seeing
man, divested entirely of consequential airs and
assumed superiority, endowed in a very remarkable
degree with common sense; being afflicted with deaf-
ness, his manner sometimes appeared brusque, owing
to his prompt and decisive answers. He was a decided
Liberal in politics, and a truly good, kind-hearted
gentleman. Mr. Foster, Mr. John Barker, and Mr.
George Jones established the great Chillington works,
in which he remained a partner up to the day of his
death, which took place at Stourton Castle, loved and
regretted by all, who, like ourselves, had the pleasure
of knowing his goodness of heart. The works and the
real estates, which were very extensive, were left to Mr.
W. O. Foster, M.P., his nephew, who has carried
them on since. Their brand of Iron has seldom been
surpassed in quality by any other firm in the
Black Country, and sells in the market at higher
prices than some other competitors' in the same
district. Mr. W. O. Foster, besides the mills and
forges, has Blast Furnace establishments at Shutt End
and Madeley Court, in Shropshire, and malleable Iron
works at Brockmoor and Brierly Hill, in addition to
the parent establishment at Stourbridge. As may be
supposed, Mr. Foster left a very large property, even
for an Ironmaster, in real estate, and the great Iron
producing establishments belonging to him in Shropshire, Worcestershire, and Staffordshire, the fee
simple of which all descended to his nephew, Mr. W.
O. Foster, M.P.

About this time Mr. Philip Williams erected the Wednesbury Oak Works, three furnaces, and the extensive mills and forges, as they now stand.[1] When the works were erected they were, without doubt, the most complete in South Staffordshire. At the close of the 18th century, John and Edward Bagnall, natives of Broseley in Shropshire, who were extensive and successful mine owners, commenced at the Gold's Hill Iron Works.[2] Mr. Edward Bagnall died, Mr. John took his sons into partnership—thus commenced and progressed the great and honoured establishment of John Bagnall and Sons, which has retained its name and fame for its brand of Iron unattenuated to the present day. Mr. James Bagnall, a polite, urbane, and always kindhearted gentleman, well known to us, the late proprietor, died one or two years since, leaving the works and goodwill to Mr. Richard, his youngest brother, Messrs. Joseph and William Naylor being left trustees and managers, with a large interest in the company, and as one of these gentlemen had for half a century mainly conducted this prosperous business, the same care continues to be exercised in the manufactories which brought the Iron into such deserved repute in the market. Messrs. George and Edward Thorneycroft,

[1] The Grand Père of the gentlemen who still carry on these great works, and continue to make the well-known brand of Iron called the 'Mitre.'

[2] Having mines and collieries at Wednesbury, Darlaston, and West Bromwich. At a later period, John Bagnall and Sons added the Birmingham Coal Hill Company's Works, The Capponfields, Bentley and Groveland Properties, with the Imperial Works, at Wednesbury. We believe the Messrs. Naylor have an interest in these works with Mr. Richard Bagnall.

after being well qualified by practical experience at Moorcroft, under Mr. Addenbrooke, commenced the Shrubbery Works at Wolverhampton. The practical knowledge and untiring perseverance of the firm, soon brought their Iron under the notice of machinists and engineers; its quality was highly appreciated and eagerly sought after by this class of consumers. Mr. Edward being constantly in the works, Mr. George Benjamin from the first ably managed the commercial department. Their Bullet Iron became an article of celebrity in Manchester for certain work in the cotton machinery, and was in great request, at highly remunerative prices. These works have been enlarged from time to time, and for a long period (fifteen years since) the Old Bradley Works, formerly occupied by the late John Wilkinson, were carried on by this firm. They subsequently purchased the Swan Garden Iron Works, which have been considerably enlarged. These, with the Shrubbery, constitute their present malleable Iron works, in all seventy-four puddling furnaces. This firm has a good name for plates and best Iron of all kinds, convenience for rolling the largest plates in Staffordshire, with reversing gear; they have likewise facilities for making angles of great length, and large sized rounds and squares. The present partners are Major Thorneycroft (the major is a Conservative, still quite the most popular man in this firm with the Wolverhampton people, the constituency being Liberal to the back bone), Tettenhall; Mr. John Hartley, Tongue Castle; Mr. John Perks, Slade Hill; Mr. Thomas Castevens, The Birches. They have two Blast Furnaces

[Margin notes:] George Benjamin. Thorneycroft, founder of the firm of G. B. Thorneycroft & Co.

Major Thorneycroft, only son of Geo. Benjamin. Mr. John Hartley, son-in-law, is managing partner.

at Bradley on a large colliery purchased by the firm, twelve or fourteen years since, which contains thick coal, &c., &c. The late Mr. George Benjamin Thorneycroft—the first mayor of Wolverhampton, whose name and fame can never be forgotten in the Iron trade—called his only son, the major, and Mr. Hartley, the surviving partners, to his deathbed, and earnestly urged them, by all means to forecast for a supply of coal for the works in the future, evidently foreseeing the scarcity which might soon be felt in the Black Country. The price to-day in London for house coal is 50s. per ton![1] This last advice was given with great earnestness.

The Great Chillington Iron Company. The Chillington Iron Company was the next large concern established. Three gentlemen, Mr. James Foster, of Stourbridge, Mr. George Jones, and Mr. John Barker, of Wolverhampton, entered into partnership in 1822, and leased from Mr. Giffard, of Chillington Park, 110 acres of land, within a mile of Wolverhampton. Here they found some of the richest mines of coal and iron in the county of Stafford, and built four Blast Furnaces with forges and mills for the manufacture of all descriptions of finished Iron. The works were erected after the designs and under the superintendence of Mr. John U. Rastrick, a young man who subsequently attained great eminence as a civil engineer, in the railway world. When completed, they were a model in construction and arrangement for that period. The management devolved on Mr. Barker, who raised their reputation to a high standard, especially as re-

[1] January 1873.

gards slit nail rods, and rails. The firm subsequently built three more Blast Furnaces at Moseley Hall, and acquired the Leabrook and Capponfield Works.

The year before his death, Mr. Barker was appointed High Sheriff for the County, being the first Ironmaster permitted to acquire a position hitherto confined to the landed gentry. His two sons, Mr. George Barker (the present chairman of the Iron Trade), and Mr. Thomas Barker, purchased the interests of the surviving partners, and still further extended the capabilities of the concern, by leasing 200 acres of mines and Blast Furnaces, under the Earl of Lichfield, at Bentley, near Walsall.

At the commencement of 1872, proposals were made to them to transfer their properties to a Joint Stock Company, which they accepted, Messrs. Barker remaining for the present as Managing Directors, and earning handsome dividends for the shareholders.

W. and J. S. Sparrow, at an early period, had William Bilston Mill. Mr. J. S. Sparrow died, upon which an Hanbury arrangement was made, which closed the partnership; Sparrow. but subsequently, Mr. John, the son of the said deceased, joined, and remained a partner up to his uncle's death. Mr. John still carries on the Bilston Mill, where rods, bars, and hoops, are turned out of good quality; and the Stowheath furnaces, which so much contributed to the colossal fortune made by the late William Hanbury Sparrow, who died worth from £1,300,000 to £1,500,000. These works are still carried on under the style and firm of W. and J. S. Sparrow. Mr. William Hanbury Sparrow was looked up to particularly during the latter part of his life,

A PAIR OF THE EARL OF DUDLEY'S THICK COAL-PITS IN THE BLACK COUNTRY

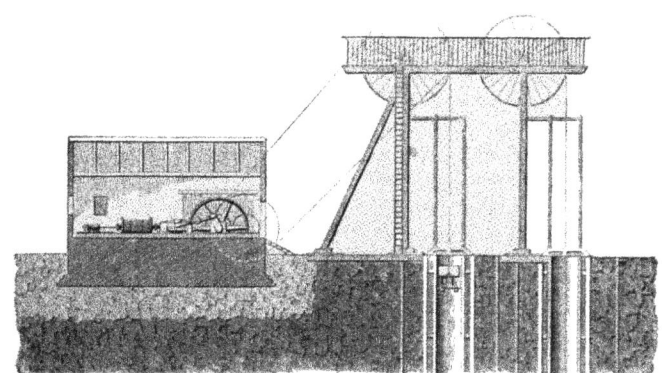

A PAIR OF THE EARL OF DUDLEY'S COAL PITS.

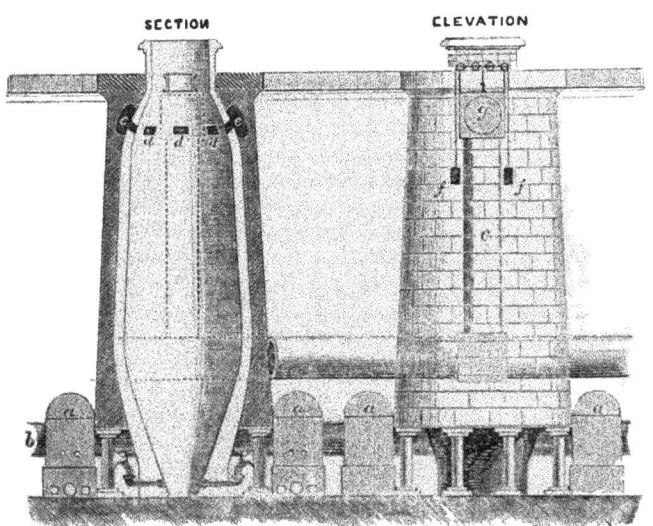

SECTION AND PLAN OF MODERN BLAST FURNACES.

with love and respect by worthy parties of standing, for advice. He was discreet, cautious, plain in his manner, with an abundance of common sense. He established the Bilston District Banking Company, and lived to see it prosper, and become one of the best managed and safest banks in the district.

The New British Iron Company are an old leading house, having Brierly Hill and Corngreaves Works, with sixty-four puddling furnaces, and six Blast Furnaces at the former works. The Lion brand of Iron is well known and appreciated throughout the country. This Iron has always commanded the best markets at high prices, being known at all the smithies and engine shops of Europe. Addenbrooke, Smith & Pidcock, are the descendants of the oldest Ironmasters in Staffordshire (the Addenbrookes we refer to). The Addenbrookes worked the Moorcroft concern in Mr. Wilkinson's time ; they have three Blast Furnaces at Rough Hay, and have obtained a patent for drawing off the gas which supersedes all others. We saw it in operation at the Dalmellington Works, in Scotland, the other day, with pride and admiration. Addenbrooke, Smith, and Pidcock, and Wm. Ward and Sons, of New Priestfields, have always made first-class forge Iron in the district, and invariably get a top price. *The New British Iron Company.*

Messrs. Addenbrooke, the descendants of the oldest firm in Staffordshire.

The Earl of Dudley's Malleable Iron Works, are situated at Round Oak, Brierly Hill. They were planned and erected by the late Richard Smith, who, up to a short time before his death, was agent to the noble Earl. We have frequently walked through them and admired the arrangements from beginning to end. The works stand *The Earl of Dudley's Round Oak Works.*

EARL OF DUDLEY'S ROUND OAK IRON WORKS.

on an extensive plateau, close to the main Stourbridge road, the Stourbridge canal ministering most conveniently to the whole west side; a public railway, likewise, serves up to the works; the arrangements for loading and unloading coal, Iron, and all other materials, are perfect, securing the very *minimum* labour cost for the manual power exerted. There are fifty-four puddling furnaces, with all appliances of the most perfect description, adapted to convert the puddled Iron, made here into bars, rods, and hoops. Angle Iron of all sizes, lengths, and shapes, small bridge rails, T, and other kinds, turntable iron, crate bar iron, nut and bolt, boat beam, boat bead convex, and indeed all kinds of shapes and forms, too numerous to be mentioned. An enormous investment has been made in rolls alone. The most difficult forms and fashions are rolled at these works, which may with truth be called the model works of the Black Country. The excellence of the Iron is acknowledged and stands unrivalled in the market, and at this moment is fetching readily 12*s.* 6*d.* per ton over the ordinary[1] brands of the leading makers in South Staffordshire.[2] The most surprising circumstance in connection with Round Oak is the promptness with which consumers of best Iron have endorsed their approval of his Lordship's Round Oak brand, the works being of recent date compared with the S. C. Crown and others. We believe the extraordinary quality of this

[1] We must except the B.B.H. brand which always follows the Earl's price within 12*s.* 6*d.*, but it must be remembered that the price of B.B.H. is quoted at the works.

[2] The Earl's price to-day (January 1873) for Bars is 16*l.* 12*s.* 6*d.* per ton; last July it was 16*l.* 12*s.* 6*d.* per ton.

EARL OF DUDLEY'S ROUND OAK IRON WORKS.

Iron may be attributed to four prominent causes; but before giving them, we must premise the explanation by stating that his Lordship has still great stores in the crust of his large domains, of all kinds of the best argillaceous ironstone in the Black Country, and his thick coal in this Eldorado is inexhaustible. In the first place, the Earl has complete machinery for manipulation. In the second place, the best men are secured for the work. In the third place, all the Pig Iron used here is smelted and made at his own furnaces—with the pick of his own mines. In the fourth place, selection is made of his best sulphurless thick coal, for puddling and mill furnaces. Nothing but 'bull-dog' is used for fettling. Fifthly, the management, under the supervision of Mr. Fisher Smith, is perfect, Mr. William Casson, Mr. Smith's able and talented deputy, being always present at the works.

Under all these circumstances, the Earl of Dudley's Iron takes its proper position in the market, which with Bloomfield B.B.H., is the top of all Staffordshire makes, and will relatively, though perhaps gradually, continue to ascend in value, while the present favourable circumstances continue to facilitate the desirable results aimed at by the astute and able management which has always directed these works.

It would be impossible to write a chapter on Staffordshire without referring to a few of those old Ironmasters, called the leading houses; those above mentioned all belong to this class. There are numerous other makers, whose brands of Iron stand high in the market; it would be impossible for us to notice all of them in these

W. BARROWS & SONS' BLOOMFIELD IRON WORKS TIPTON STAFFORDSHIRE.

pages. Our report of the quality of those mentioned may be implicitly relied upon, and with a notice of the last, and the most important of them all, we must close these sketches, referring our readers to the general list, which contains the name and particulars of all those works embodied in the furnaces, mills, and forges of the Black Country. W. Barrows and Sons, proprietors of the Bloomfield, Factory,[1] and Tipton, Iron Works, have one hundred puddling furnaces, quite the largest number of any iron works in South Staffordshire, and produce 1,000 tons of finished Iron per week. The brand is B.B.H., taken after the names of the originators of the firm, Bradley, Barrows and Hall, and is known and appreciated, not only in this country and the Colonies ; but its world-wide fame has caused it to be eagerly sought after in all parts of the world, being preferred in Australia and other Colonies before all others. It has for many years deservedly been the most favourite English brand imported into Holland and the Low Countries, the excellency and uniformity of its quality having created a living faith in the B.B.H. brand which *no other Iron made can boast of to the same extent*, in the number of its customers, or degree of implicit faith willingly reposed in the quality of the article itself.[2] The Bloomfield Works were erected by the above-named gentlemen a little later than 1826, having from time to time been amplified to their present dimensions. The brand remains unchanged,

---

[1] See illustrations.

[2] Mr. Chance told us that the iron made here was so valuable for certain glass-making purposes as to be worth its weight of silver.

W. BARROWS & SONS TIPTON GREEN IRON WORKS.

W. BARROWS & SONS TIPTON GREEN IRON WORKS.

W. BARROWS & SONS FACTORY WORKS. IPTON.

F

Cinder pig iron or inferior mine pig iron is never used here.

the quality, amid the changes of fifty years, has remained the same, unattenuated, unadulterated, always uniform, B.B.H. in quality. All the old partners have been long since dead and gone to rest, still Mr. Hall's original plan is carried out, both in the kind of pigs used, the boiling process which he invented, the best workmen engaged, and perhaps better pigs and coals used than anywhere else, with good managers on the ground day and night. This large concern belongs to the late Mr. William Barrows's sons, Colonel Barrows, and Mr. Joseph Barrows, being the managing partners, and follow their father's footsteps in attending daily at the works. The superior quality of the Earl of Dudley's pig Iron is well known in Staffordshire, in fact, it is quite the best. The name and fame of Earl Granville's Lilleshall is equally celebrated. W. Barrows and Sons purchase and use more than half the quantity of Earl Dudley's Pig disposed of in Staffordshire, and quite half the whole quantity of Shropshire Pig sent into Staffordshire, is consumed by this firm. They still adhere to their father's plan of buying none but the very best; it is therefore rigid attention to this rule, which enables the present firm to make such very uniform and matchless Iron in quality, which is seldom equalled and never surpassed by any other house. In fact, the uniformity of the quality of B.B.H. Iron is not equalled by any other house, and their quality remains unrivalled.

William Dawes and Sons are the proprietors of the Bromford Iron Works. These are large works very near to the Oldbury station on the north side of the Stour Valley Railway, and can be distinguished by the round chimneys or stacks, the diameter increasing as the stack rises, the summits of which are considerably wider than the base. This is a very old and highly respectable concern. The family are all born gentlemen, and educated Ironmasters, make and export nail rods largely, particularly to China, where they are highly esteemed for quality. 'Dawes and Sons' is branded on them in Chinese characters. Mr. George Dawes has the Elsecar Works in Lancashire, and W. Dawes and Sons, besides the Bromford, have likewise two Blast Furnaces at Withymoor, near Dudley. The make of Pigs here is first class. Dawes and Sons, of Bromford, stand well in the market for rods, hoops, and bars; the Iron is soft and good, their rods and hoops are exported largely by the Liverpool merchants; they make bars, hoops, rods and small and large angles, small and large rounds and squares, sheets and plates. These works have always been carried on very regularly, and the quality of the Iron is noted for its uniformity. The firm is well supplied with coal of the best quality from its own mines, and the 400 tons of high-class Pig Iron, made at their Withymoor furnaces per week, gives a thoroughly Staffordshire character to the malleable Iron they produce.

Having noticed all the most famous malleable Iron makers, before concluding this part of Staffordshire metallurgical manufacture, we must refer to the Steel manufacture, which, although not so extensive here as in Sheffield, and at Barrow-in-Furness, there are one or two concerns which claim particular attention. Twenty years since, a method of puddling Iron into Steel was discovered in Staffordshire, which, for

MINERVA IRON & STEEL WORKS.

BEAVER IRON, STEEL & SPRING WORKS.

certain kinds of Steel, was admirably adapted to the wants of the Sheffield converters ; Messrs. Solley, of Leabrook, were the first to make it.

The Chillington Company went into this trade, and were very successful in producing the article. At the same time, Mr. Isaac Jenks commenced making this article at the Minerva Works, and also since, at the Beaver Works at Wolverhampton (both of which we give an illustration of), having previously been instructed in the manufacture at Messrs. Solley's. Mr. Jenks's practical knowledge of this trade enabled him to carry the Minerva brand into all the great markets. Through his constant attention and practical knowledge of the manufacture, the Minerva brand stands unrivalled. From the commencement, the fame of Jenks's puddled Steel, so admirably adapted to the manufacture of railway springs and other purposes, crossed the Atlantic, and the demand in the great Republic has been considerably greater for this brand than any other. Four-fifths of our export of this kind of Steel to the United States last year was manufactured at the Minerva. These works have grown, *pari passu*, with the demand, and no doubt the proprietors' wealth in the same ratio, and now make all kinds of the best Steel, which is preferred in America to the Sheffield makes, for while Iron was unsaleable and low in price, the demand for Jenks's Steel continued to increase for America, which brought relatively high prices, and, no doubt, fair profits. And although he commenced with slender means, the first in this speciality, he is now, without doubt, one of the most wealthy Ironmasters in Wolverhampton. Mr. Jenks is a staunch Wesleyan Methodist, very quiet, and unassuming in his intercourse with others, unexalted in his own estimation by his accumulation of wealth. He was unanimously appointed Mayor by the Corporation of Wolverhampton, upon which the Superintendent Minister was made the Mayor's Chaplain, and the Corporation attended Darlington Street Chapel, with their Mayor, to hear a Methodist sermon, in their Corporate capacity. Mr. George Benjamin Thorneycroft was the first, Mr. Jenks is the third Ironmaster who has been Mayor of the Capital of the Iron Trade in the Black Country.

Messrs. William Hunt and Sons, of the Brades, are the oldest Steel manufacturers in Staffordshire. This is a very old and highly respectable concern, and the article turned out is of excellent quality. They have seven puddling furnaces and three mills, with extensive converting conveniences on the old Sheffield plan ; cast, shear, blister, and all other kinds of Steel are made here, of the best quality ; quality, not quantity, being the great object of the proprietors. The works will be found on the left of the Dudley Road, near to the devoted town of Oldbury.

The Darlaston Iron and Steel Company took to Bills and Mills' old establishment, which had a first-rate connection for Iron and Steel ; the last partner in the old firm retired with between two and three hundred thousand pounds ; the Lloyds, of Wednesbury, have a large interest in these works, which are situated at Darlaston Green and King's Hill ; they have thirty-eight puddling furnaces and eight mills. They have rich and abundant coal mines, all the best ironstone measures, and in addition to their own supply sell coal largely. This firm has three blast furnaces. This is a Limited Company, which has been successful in dividends to the present comparatively new proprietors. All kinds of Iron and Steel are made here of the very best quality. Mr. Thomas Wells, of the Moxley Iron Works, has twenty-two puddling furnaces, three bar and plate mills, makes good plates, and manufactures Steel.

There are other makers of Steel in Staffordshire, particularly puddled Steel, mostly in the neighbourhoods of Leabrook and Oldbury, Solley Brothers, a Limited Company, being the oldest and largest of this class.

There is another Staffordshire speciality in Iron making, which would be understood better if we say Marshall and Mills, of the Monway Works, Wednesbury. These gentlemen, many years since, succeeded in making the best gun-barrel Iron in the world, and have for years supplied the Iron to the Birmingham gun makers, also to the British and American Governments, the quality of their Iron being approved and sanctioned by both Governments. The price they obtain now would be about £33 10s. per ton. [1] The firm was dissolved, by mutual consent some years since, Mr. Mills having grown-up sons whom he wished to introduce. Mr. John Marshall remained at the

---

[1] They make sheets and other qualities, commencing bars at 18l. and sheets at 24l., but their best quality sheets fetch much higher prices; their best best bars 42l. per ton.

Monway Works, which has eleven puddling furnaces and two charcoal fires, and carries on the original business. Henry Mills and Sons erected the Victoria Works at Walsall, a modern unique concern, which Mr. Mills carries on in connection with his sons, making also, the same specialities which gave so much celebrity to the firm of Marshall and Mills. Mr. Henry Mills took his first lessons in Iron making from the late *grand père*, Mr. Phillip Williams and also Mr. John Bagnall. He was on intimate terms with our own dear friend, Mr. Hall, of Bloomfield, whom we have often heard say that 'Mr. Henry Mills knew how to make Iron.'

Since Mr. Hall's death, Mr. Henry Mills, as a practical and scientific Ironmaster, has had no rival in South Staffordshire. He is a staunch Methodist, and often boasts that his father lived and died one also. Both Mr. Marshall and Mr. Mills make a superior kind of sheet Iron, which realises a very high price in the market. The Victoria Works has seven puddling furnaces only, quality rather than quantity being the object aimed at by Henry Mills and Sons.

LLOYDS, FOSTER & CO.

The Old Park Works, Lloyds, Foster and Co., Wednesbury, is one of the most time-honoured and important in the district, and was erected by Samuel Lloyd, who was afterwards joined by Mr. Foster. There are three blast furnaces, and most extensive engine machine shops and foundries, with lathes and

machines quite abreast with the shops at Manchester
and the Tyne. This wealthy old firm built our beautiful
new Blackfriars Bridge. The mines of the Old Park
produced a quality of Pig Iron of the very highest
class, and the high position, wealth, and known integrity of the firm, placed within their reach orders for
bridges and other public works, which contributed to
the wealth and celebrity of this time-honoured establishment. The proprietors were mostly 'Friends;' the
late Samuel Lloyd was managing partner, with the
assistance of Sampson Lloyd, during the extent of our
recollection. Samuel Lloyd was never excited; firm,
urbane, considerate, and kind to those beneath him :
always the same. In the large transactions we have
from time to time carried out at this office with Samuel
Lloyd, his yea was yea, his nay was nay; he was
always himself, always the same, we never heard of
any bickering and disputing in respect to any business
transaction of Samuel Lloyd's. The number of men
employed here, in all departments, perhaps exceeds
that of any other concern in the Black Country. The
paternal kindness of Samuel Lloyd endeared him to
the men—all artisans and workmen thought themselves
fortunate to get a job, as they said, at 'Quaker Lloyd's.'
The greatest kindness was shown to the widows and
orphans of those workmen of this firm who lost their
lives in their collieries; schools were erected at their
expense, in which these orphans were educated, and
when Samuel Lloyd died, their own workmen lamented
his death, and the universal remark was, 'That a good
man had passed away.'

This firm kept a large truck shop, which, like the works, was conducted on high moral principles; all articles were of the best quality, the prices were quite as low, and sometimes lower, than the shops. Samuel Lloyd took pride in buying the chief articles himself, particularly the tea, the bullocks and sheep; the shop was noted for the best butcher's meat in Wednesbury. He often walked up and looked over it. We accompanied him on one occasion, and were gratified to see how minutely this great man could stoop to little things, when he knew that his labour was for the good of his workmen.

The Old Park Company, seven years since, was merged into a Limited Company, called the Patent Shaft and Axle Tree Company, which, combined, is one of the largest concerns in Staffordshire, having three Blast Furnaces and eighty-six puddling furnaces, with the great foundries and fitting shops above referred to. The Patent Shaft and Axle Tree Company Limited, is one of the most prosperous and paying concerns in England, and, we believe, will always continue so. The Lloyd family have still a large interest in it.

Wednesbury is the seat of the gas-tube trade, which has become, during the last fifty years, so important for the supply of the home and foreign markets of this ingeniously made article. James Russell and Sons are the original patentees, and continue the largest and most celebrated makers, exporting their tubes in increasing quantities year by year to the great Republic, Russia, France, Germany, and other markets, where the name is so well known. This great concern is now

GEORGE ADAMS & SON, NEW PRIESTFIELDS IRONWORKS.

a limited
Mr. W. S
agent of th
manageme
prosperou
prietors."

Messrs.

establishe
machinery
and is fa
engines, p
for Iron
foundry
situated t
for a cen
bridges, s
Iron. M

MARS' L

These
carried c
style an
graving
expressl
of Great

The

Mr. S
James
taken i
engrav

a limited company, under the able management of Mr. W. Smith,[1] who is brother to Mr. Fisher Smith, agent of the Earl of Dudley. Under Mr. William Smith's management, this valuable old concern has been highly prosperous, returning large dividends to the proprietors.[2]

Messrs. T. and J. Roberts's great Swan foundry, established 1824, long celebrated for pipes and machinery, is situated near here at Westbromwich, and is famed for all kinds of castings for steam-engines, pipes, soft and chilled rolls, and all other rolls for Iron works, where a great stroke in the general foundry business is got through. Not far distant is situated the well-known Horsley Company, established for a century or more, famed for the manufacture of bridges, steam-engines, and all other constructions in Iron. Mr. Broad is the managing partner.

MARS' IRON WORKS, PRIESTFIELDS, WOLVERHAMPTON.

These works were erected by, and have since been carried on by, Mr. George Adams, recently under the style and firm of George Adams & Co. The engraving opposite is made from a photograph taken expressly for the editor of the 'Guide to the Iron Trade of Great Britain.'

The quality of the iron made here is first-class,

[1] Mr. Smith is Managing Director and Chairman of the Company.
[2] James Russell & Sons are the oldest and most extensive gas-tube makers in the United Kingdom. We believe them to have been the original inventors and patentees, and still take the lead in the trade.

Mr. Adams from boyhood having been practically connected with the manufacture of iron in Staffordshire, first as manager for Rose, Higgins & Rose of the Bradeley Field Works, whose plates and sheets under Mr. Adams' exclusive management acquired considerable celebrity in the London market. At a more recent period, say sixteen years since, he undertook the management of Wright & North's Monmoor Works, which remained under his control in all departments for ten or twelve years, during which the character of the Monmoor Works' brand for plates and sheets was raised considerably in the home markets, where their plates are well known. Mr. Adams then built the works engraved above, and has, during the last seven years, successfully carried them on.

The sheets, hoops, plates, bars, strip and small rounds and squares manufactured here, may be relied upon, Mr. Adams's eldest son being always at the works, under the constant supervision of his father whom we venture to say, as a practical man experienced in practical iron-making, is rarely equalled in South Staffordshire. We believe Mr. Adams still supervises the extensive Monmoor works, which, since the retirement of Mr. David North with a large fortune, are carried on by E. T. Wright, Esq. The Monmoor brand of Iron is well known and highly appreciated.

Any of our London friends who order iron from George Adams & Co. will be satisfied with the quality. The firm are particular in the raw material they use, and for a mixture purchase both the Earl of Dudley's and the noble Earl Granville's Lilleshall pig-iron. The works are well situated, being on the banks of the Birmingham Canal, with a small arm or wharf which runs into and intersects the works. The situation is likewise favourable for coal, pig-iron, and labour, being midway between the Bilston and Wol verhampton markets. The works, being of recent erection, are well adapted to turn out large quantities of iron, from being equipped with modern machinery abreast with the progress continually being made in metallurgical establishments of this kind.

CHAPTER VI.

BILSTON: ITS VALUABLE COAL MINES AND IRONSTONE.

WE must crave permission to say a word about Bilston, our native township, before we close these sketches. The whole of Bilston is built upon a long slang of the crust of the earth, overlying the most valuable coal and Iron mines, taken altogether, in the world, the ten yard thick coal being better here than any other part of this wonderful deposit. Besides this, there is the new mine and fire-clay, the Bruch, Flying Reds, the Heathen coal, and the Bottom coal.

With regard to Ironstone, all the best measures are there, including the new mine, or Whitestone, Balls, Blue Flats, and Poor Robbins, all in their Bilston prime; but these valuable measures have been gotten all round the town, and, either by fair or foul means, taken from under the outer portions of several parts of Bilston proper: they remain, however, in High Street, which is the longest street here, and about one acre occupied by St. Leonard's Church, the Swan bank, and the graveyards. Bilston Church and its churchyard proper, with the parsonage, occupies about one acre of land which contains the coal and mine, with all the usual measures in the virgin state, the fabulous value of which is incredible. To give the reader an idea of the importance and value of the coal and iron mines of our native town, we give a valuation of the same seriatim below, being the price of the mines and minerals brought to grass, which foots up to no less than £41,440 sterling for the coal and Iron mine of a single acre of land, all the mines being deposited in their Bilston prime here. We have valued the materials at the price of to-day (January 1873). As this thick coal and the same mines underlie the Bilston district, the reader can form his own opinion of the mineral treasures contained therein. Our valuation has been submitted to, and endorsed by, mining authorities in the district, and is as follows :—

For the thick coal, the Heathen coal, New Mine coal, the Friesley coal, the Bottom coal, and Mealy Grey coal, are all there, and would yield in the aggregate 56,000 tons ; taking coal and slack altogether at 12s., much below the present value, per ton would give a value of £33,600. The Ironstone measures known to be there are the New Mine, Poor Robbins, Balls, and Blue Flats : these would aggregate 6,400 tons. Taking the average value of these Ironstones at 22s. would yield £7,840, or a grand total of £41,440 per acre ; or for the acre and a half of coal and iron which belongs to the church, £62,160. This would be the gross value of the minerals and coal if they turned out well when brought to grass. From this, however, deductions must be made for Royalties, sinking, engines and machinery, labour, interest of capital, &c., &c., &c. We do not say that even at this price it would pay to make the necessary outlay for so small a piece of land. We have given this simply to show at one view the great value of mineral property in this neighbourhood.

THE PATENT NUT AND BOLT COMPANY, LIMITED.

The works of this Company are worthy of special notice, as being the most extensive in the United Kingdom for the manufacture of the immense variety of articles coming under the denomination of bolts and nuts.

The Company's principal establishments are at the London Works, near Birmingham (which is also the head office of the Company), the Stour Valley Works, Westbromwich, and at the Cwm Bran Works, near Newport, South Wales. A notice of the latter will be found on the following page.

At the London Works are manufactured all kinds of engineers' black and bright bolts and nuts, coach screws, rivets, and washers, also every description of bolts and nuts used by ship builders, agricultural implement makers, telegraph engineers, railway carriage and wagon builders, &c.

The Stour Valley Works are devoted exclusively to the manufacture of railway fastenings such as fish bolts, fang bolts, and spikes in all their endless variety, and it may be observed that the whole of these articles are made from Iron puddled and rolled on the premises. The works also contain a foundry for casting railway chairs.

Mr. Arthur Keen is the Managing Director of the works in Staffordshire, and Mr. Edwin J. Grice of the works, collieries, and blast furnaces, in South Wales. Mr. W. F. Jones is the Secretary of the Company, which was incorporated in the year 1864.

CWM BRAN WORKS IN WALES.

The Cwm Bran Iron Works and Rolling Mills, near Newport, are the property of the Patent Nut and Bolt

Company, Limited, and are engaged exclusively in the manufacture of all the various kinds of railway fastenings, including fish plates and sole plates.

Attached to these works are extensive collieries and also blast furnaces, the latter producing the well known ' Cwm Bran' pig Iron, a brand which has always been highly valued, more especially for the manufacture of shot and shell.

Being at Smethwick a few months since, we were permitted to walk through the London Works. We were astounded at the facility with which the nuts and bolts were forged, and the screws cut and turned out bright and beautiful.

The nuts are made stronger and better than the hand of man could forge them. By clever machinery, the Iron is concentrated and rendered more dense and resisting, and we feel safe in saying that no forged nut can compare with them. The manufacture of the bolts was just as marvellous, as rapid, and as perfect as the excellence of the nut and screwing machines was admirable. We, however, observed one thing particularly. The Iron was uniform and of very high quality, for in the thousands of nuts and bolts that we saw made, after examining, we could not discover a single rent in punching of the nut, or crack in the heading of a bolt—the Iron pressed in the machine like plastic potter's clay.

The nuts and bolts are made with square, hexagon, or octagon shaped heads and nuts in the very highest style of workmanship. It was a great treat to us to have the privilege of seeing the machinery of these works.

CHAPTER VII.

BILSTON : ITS MANUFACTURES.

Numbers of females are employed in japan painting here.THE manufactories at Bilston are not numerous; the great staple is japannery and tin plate wares, Mr. Farmer and Smith and Edrington, of the New Town, and Messrs. Jones and Rowley, of the Old Town, being now the principal makers. The japan and tin plate wares of Bilston are not of the very highest class; our Bilston friends make some very good articles, but as labour is cheaper here than anywhere else, the manufactories do a large business with the Birmingham merchants, who run the prices down when they get a chance during a bad time of trade. On this account the Bilston manufacturers have to produce an article at a lower cost to meet the Birmingham market. There is a good market on a Saturday night here, attended by colliers, puddlers, shinglers and rollers, with their wives; a large quantity of ducks, geese, and fowls are in constant demand by the Ironworkers; good beef and pork find a ready sale among these stalwart artisans, who descend in swarms from the smoky regions of Bradley, Shropshire Row, and Hallfields, to spend their money on a Saturday night.

We must, however, say that the colliers are a very quiet and orderly body of men; they may occasionally take a little too much to drink on a Saturday night.

The Ironworkers are intelligent, respectable, and well behaved, and move much higher in the general social scale than any working men we know in any part of England. When the last sixteen weeks' strike of the colliers took place, we called them together in Bilston, and from the window of the Shakespeare Inn addressed the largest meeting of them ever assembled in Bilston, on the Saturday, advising them " to abandon the strike and go to work ; the trade was so bad in Iron, that there was no hope for them "; they took our advice, and went to work on the Tuesday following. At the great lock-out of the Ironworkers in 1862, which lasted eighteen weeks, when all efforts had failed at reconciliation, the Earl of Lichfield finally endeavoured to bring about a settlement, but without effect. Impelled by a sense of gratitude, which the Bilston colliers and Ironworkers had imposed upon us, by giving us a majority of their votes at the Parliamentary borough of Wolverhampton election, which we had so recently contested in the Liberal interest, against Mr. Weguelin, the present sitting member, and Mr. Staveley Hill, now the well-known Queen's Counsel and member for Coventry, we issued a placard, calling a meeting of the puddlers and Ironworkers; and after addressing the largest meeting ever held during the eighteen weeks' lock-out, we proposed three resolutions, which the men passed unanimously by a show of hands, which induced the masters to open the gates of their works, and the men went in on the following Tuesday and Wednes-

We were second on the poll, and not far behind Mr. Weguelin.

day, on the old terms.   On this occasion, from the first, we advocated the working man's cause, believing as we did, that the masters were wrong when they locked out the men.   From that time to this, now some ten years past, we are happy to say that, with very little interruption, peace, harmony and goodwill have existed between the masters and the men, although the masters have had hard times of it during eight of the ten years which have intervened.

Bilston is surrounded on all sides by Ironworks, collieries, Iron foundries, and coal mines.   The famous Iron foundry of T. Perry and Sons, of Highfields, is near here, where steam-engines, chilled and soft rolls, and everything appertaining to an Iron works is made. The firm is famed particularly for blast-engines,[1] all heavy castings, likewise the very best safes, which are well known in the London market for their excellence and superiority over other makers.

The late Mr. T. Perry was a director of one of the principal banks; he was a man of considerable influence and of unspotted reputation.   Mr. Thomas, his second son, is managing partner.

Messrs. Thompson and Hatton's[2] Tin-plate works are situated here.   Groucott's, Bradley Bridge ; Messrs. Hampton, Brierton and Cole, the Bilston Sheet Iron Company,[3] George Hickman's works, Mr. Alfred Hick-

---

[1] The first blast engines which we saw started at the Barrow Steel Works were built by T. Perry & Sons, which are so much approved of, that they have retained the appointment, and since built all the blast engines for this great establishment.

[2] Mr. Hatton the banker at the Joint Stock Bank.

Chambers and Sankey make a larger quantity of Sheet Iron than any other Ironmasters at Bilston, their quality is very good, their

man's furnaces, and Mr. G. Merriman's Lanesfield Iron Works are all in a group, beneath the curtain of black smoke which forms the normal canopy of Bilston. Here too the Iron works of W. and S. Sparrow are situated, one of the oldest and most wealthy concerns in the Black Country. Turleys' and Fowler's blast furnaces, and also the famous Capponfield furnaces, belonging to John Bagnall and Sons, emit their smoke and flame, and produce Iron of their well-known brands. All the above works are within the radius of the Bilston group.

One hundred years ago, the principal manufactures of Bilston were Iron and silver buckles, steel watch chains, snuff and tobacco-boxes, and enamelling designs in copper. Bilston, like Wednesbury, was then well stored with fighting cocks and bull dogs. Mains of cocks were fought, and at the Bilston wakes as many as three bulls were frequently baited while these cruel sports were permitted by the Government.

The cocks were kept out at walk in the surrounding districts by those who indulged in this unfeeling sport. Bull dogs were then kept by numbers of the colliers to bait the bulls with at the wakes, Catchem's-Corner and Hell Lane, a place no longer known by the appellation, or by the recurrence of this brutal sport.

Thank God! things are much changed since then. Two new churches have been erected. A large Baptist church exists here, a splendid Methodist chapel,

doubles and latten always fetch a good price in the London market. The firm is called the 'Bilston Sheet Iron Company.' Works near to a railway station.

a New Connection chapel, a very fine one, one of the best Primitive Methodist chapels in the district, besides the old St. Leonard's Church, St. Mary's, and one Independent chapel.

In 1872 the cholera raged more here than in any other town in England when the Rev. Mr. Leigh so successfully appealed to the kind-hearted English people for pecuniary assistance. The clergyman of St. Leonard's is invariably elected for life by the householders; and it was at Bilston, where our late dear and ever beloved friend the Rev. J. B. Owen, devoted the greatest part of his useful life at St. Mary's Church, where his memory will ever be cherished for his great ability, unrivalled tact, Christian virtues, and untiring zeal, coupled with indomitable industry in the work of his Lord and Master.

CHAPTER VIII.

THE BILSTON NECROPOLIS ON THE COAL MINE BANKS, AND WOLVERHAMPTON.

BEING now within two miles of Wolverhampton, which is considered the capital of the Black Country, we must refer to it. Leaving Bilston then by the highway, we pass the Cholera Orphanage, and on the right, observe a little farther on, the Bilston Cemetery, which, to a stranger, has a very singular appearance, although well looked after. Situated on the very top of the old mines, *Bilston Cemetery on the top of the old coal mine.* the land all round for miles is devastated and thrown into heaps. Attempts have been made to level the ground and recultivate the surface, but nature, after the torture which she has undergone, refuses to give her increase; the trees all withered away, the sun appears to frown on the efforts of man, and refuses to force his genial, sparkling rays through the murky atmosphere which enshrouds this devoted spot. At night the distant dim lights throw their ghastly flickerings over the graves and tombstones of this disembowelled necro- *Nocturnal shadows on the grave stones.* polis, rendering the place a source of solemn reflection to the midnight traveller, and of melancholy contemplation to a cluster of inhabitants who dwell in the precincts of the dead. The road and the country all round has been disembowelled in all directions, and no doubt if a correct survey could be made, the adits and gate-roads resemble the catacombs of Rome, only that the excavations here are on a more gigantic scale than those of the old Pagan City. Nothing can be seen all round for miles, as far as the eye can reach, but blast furnaces and tall chimneys, vomiting forth volumes of dense smoke which form a dark canopy, resembling in some measure a moderate London fog; fortunately the latter is quickly dispelled by the sun's rays. Here, however, he seems to have lost his power; obstruction of the smoke renders a thickened state of the atmosphere a normal condition, of the country. Close to the toll-bar are two furnaces, which have done their work, the mines being worked out, the process of demolition having proceeded to decapitation only. During their lifetime they run out a very large fortune to the proprietor, the late William Ward, Esq.

We now proceed along this hollow road to Wolverhampton, hemmed in on all sides by Iron works and collieries, and as we near the latter town, the great Chillington works are seen on the right; on the left, on the banks of the canal, will be seen Edwin Lewis' gas tube factory; also, Bayliss, Jones & Bayliss's great works, the well-known contracting merchants, of 3, Crooked Lane, Cannon Street, London, for rails, fishing plates, nuts and bolts, and all kinds of Iron work, from massive cables and anchors to half-inch chains and small screws. Close on the left side of the bridge we have Mr. F. N. Clark's great galvanizing works, one of the largest in Wolverhampton. On the banks of the canal to the right is Perks's famous old edge tool

Bayliss, Jones & Bayliss's works.

Mr. F. N. Clark's great galvanising Works.

works, and further on in consecutive succession, are the Iron works of G. B. Thorneycroft & Co., Mr. Isaac Jenks, the present Mayor's, Minerva, and Beaver Steel Works; Baldwin's & W. Sparrow's tin plate works; T. & C. Clark's famed foundry for the enamelled hollow ware, of which they are the original patentees; Mr. Thomas Bridge's foundry for the manufacture of all kinds of castings, machines, steam engines, &c., &c.; and farther on, the chemical laboratory of W. Bailey & Son, famed in London and elsewhere for the manufacture of mercurials and choice chemicals, more than any other establishment.[1] Wolverhampton is a town of considerable antiquity, and has a fine old Gothic collegiate church. The grammar school here claims the honour of educating Dr. Johnson, also the famous Dr. Abernethy and Sir William Congreve. There are three Methodist chapels, one New Connection, one Primitive Methodist, two spacious and elegant Congregational churches, and a beautiful Baptist chapel. The ministers and clergymen of all denominations here are always of a higher and more educated class than other towns in the Midlands, not even excepting Birmingham. The late Angel James of Birmingham must be excepted. A large infirmary, a Government school of art, and an orphanage built by a private individual, which honours the memory of Mr. John Lee, and largely ministers to one of the first necessities of orphan children. This

---

. [1] Mr. Bailey is the patentee of the well-known bisulphite of lime, so justly celebrated for purifying brewers' casks, and preventing decomposition in butchers' meat, and makes the bisulphite at this laboratory.

EDWARD DAVIES'S CROWN GALVANIZING WORKS.

is the great emporium for the manufacture of stock
locks, rim, mortice and cabinet locks; indeed, most

kinds of the best house, door and cupboard locks, are made here. Chubb's great lock works have always been carried on at Wolverhampton; cut nails and corkscrews, edge tools and brass foundry, gas tubes and Iron hurdles, have long been made to a great extent here. The galvanizing trade was introduced here thirty-five years since by Mr. Edward Davies of the Crown Works, and has constantly increased with the demand, until both galvanizing and galvanized manufactures produced here have become very important Wolverhampton industries.

The wood engraving opposite represents Mr. Edward Davies's Crown Galvanizing Works. We believe Mr. Davies was the first to introduce galvanizing on a large scale into Wolverhampton, in 1838. The trade has since become one of the principal staple trades of the town. Nevertheless, Mr. Davies's article has always kept a very high position in the market. The brand of the Crown Works is a crown surmounting the Staffordshire *knot*.

His best Crown sheets have always been well known and appreciated in the market, and occupy the very highest position in the United States of America and Australia.

Mr. Davies likewise galvanizes best best best, and high-class charcoal sheets. Iron is also tinned for various purposes. Galvanized Iron houses for export are made largely at Mr. Davies's factory, which is well situated both for canal and railway accommodation.

The japannery of Wolverhampton is the best in England. Mr. F. Walton's articles, produced at the Old Hall, for artistic design, elegance of shape, and excellency of workmanship, supersede all others in the United Kingdom; there are three or four other very large concerns, all producing first-class goods in their own peculiar styles; the largest and most important of these, and the greatest favourite in the London market, is Mr. Henry Loveridge, late Shoolbred and Loveridge's. All the japanners carry on the tin plate workings. The tin plate workers of Wolverhampton are likewise in very high esteem in the London market. The factors and merchants of Wolverhampton are a very important and influential class; Mr. John Morton's being the largest and most important concern of them all, for all kinds of Iron manufactures and Iron itself. The largest stock of Iron out of London is always kept here, at the great Iron warehouse of G. & W. Underhill, in Castle Street, and an astonishing business in extent is done by this firm. Certainly the largest of any Iron merchants out of London, it is the oldest concern in the Black Country, and was established by Mr. Joseph Underhill, the father of the present proprietors, at the beginning of this century. This firm, for sixty years, have been the agents for the sale of the S.C. crown Iron made by Mr. 'Foster, of Stourbridge, the old name of John Bradley and Co. still being kept up.

Wolverhampton returns two members to Parliament since 1832, always in the Liberal interest. At the last election we contested this borough in the Liberal interest against Mr. Weguelin and Mr. Staveley Hill.

Mr. Weguelin, the present sitting member, was the successful candidate. The Ironmasters meet here on Change every Wednesday afternoon, and the Wolverhampton quarter day takes place the second Wednesday in each quarter of the year. One side of the town develops to view the Black Country in earnest, the other presents a beautiful landscape, with the charming country village of Tettenhall, two miles distant from the town.

WOLVERHAMPTON

cannot boast of street architecture to compare with New Street, Birmingham, or Sackville Street, Dublin.

The Queen Street Church, however, and one of the Methodist chapels, for architectual beauty and design, will sustain favourable comparison with anything of the kind in England. The old Gothic Collegiate Church is a fine specimen worthy of the attention of the archæologist.

The other public buildings are the Iron Exchange and Town Hall: the facade of the latter is good Italian, the former a good building, but damaged by its proximity to the Collegiate Church. There is likewise a handsome school of art, and a noble infirmary. The design of the latter is bold, imposing, and effective, but devoid of all trapping and embellishments as it should be in the execution. Wolverhampton, likewise, has a splendid orphan home as good in the design of the building as it is benevolent in its object. G. & W. Underhill's monster warehouse and Henry Loveridge's Japan Works are buildings well worth seeing; the stock of Iron at the former, and the number of hands and activity in japan work at the latter, surpass anything of the kind in the United Kingdom. Wolverhampton is well supplied with banks. The Bilston District and the Wolverhampton and Staffordshire are the oldest, and do the largest business. The Midland Limited and Lloyds' Banking Company of Birmingham also have banking houses here. The merchants, factors, and tradesmen of Wolverhampton are noted for punctuality in the payment of accounts: in this respect Wolverhampton stands higher than any town in England, the cotton Lords of Manchester excepted.

THE NEPTUNE FORGE CHAIN AND ANCHOR WORKS, TIPTON GREEN, TIPTON.

This is one of the largest chain and anchor works in England, well represented by plate opposite.

The situation is admirable, being on the banks of the canal, near to the railway and in the very centre of the coal and labour market. Mr. Joseph Wright, the proprietor, is the son-in-law of Mr. Theophilus Tinsley, the well known nail factor of Dudley, and we believe the Tildsley family are still connected with these great works. Mr. Wright's chains are famous not only in the London and Liverpool markets, but throughout the world: the cables made at this celebrated forge stand unrivalled. When in the Black Country the other day, we were particular in our enquiries in respect to Wright's chains and cables at the Government testing and proof establishment at Tipton, and were there informed that the testing and proving of the cables and chains turned out at the Neptune forge completely justified the celebrity for quality which Mr. Wright's chains have in Germany, Russia, the Levant, and other foreign markets. The anchors made here are well known throughout the world the Neptune forge having turned out from time to time the best and largest made. We understand the excellency of the quality of the chains and cables is owing in a great measure to the use of B.B.H., and the Earl of Dudley's bars, which are

TELEGRAPH MESSAGES SHOULD BE SENT TO
DUDLEY RA LWAY STAT ON.

NEPTUNE FORGE, CHAIN & ANCHOR WORKS, TIPTON.

unsparingly consumed at this great factory, although the price paid is 1*l.* per ton to 1*l.* 12*s.* 6*d.* per ton more than any other Iron of this class made in Staffordshire. Messrs. Wright & Co. are now engaged on large orders for the Russian and Turkish Governments.

CHAPTER IX.

SHROPSHIRE IRON AND COAL DISTRICT.

VARIOUS circumstances during the last 150 years have constantly ministered to the growing expansion of the Iron Trade in this Kingdom. Without her valuable pig Iron, the more southern sister county could not have made such rapid and unceasing strides in metallurgical industries during the last hundred years in the Black Country.

It was discovered at the latter end of the last century that a mixture of Shropshire with Staffordshire pig, in the puddling furnace, gave increased strength and other desirable qualities to the Staffordshire finished Iron. It is, however, to the intelligence and perseverance of the late Mr. John Horton, of Prior's Lee Hall, and Mr. James Foster (S.C. ♚ brand), late of Stourton Castle; Mr. Joseph Hall, the managing partner at Bloomfield (B.B.H.); George Benjamin Thorneycroft, of the Shrubbery, John Bagnall, of Goldshill, and Henry Mills,[1] now of the Victoria Works, Walsall, that the metallurgical manipulations which combine these Pigs with Staffordshire have justly obtained a name

[1] All the Staffordshire and Shropshire worthies above-mentioned have long since ceased from their labour, except the last-named, Mr. Henry Mills, whom we were pleased to greet in good health on the last quarter day.

and fame for Snedshill, the most noted works for best quality in Shropshire, and also B.B.H. Staffordshire bars, hoops, and sheets, and wire rods in every civilized country of the world.

The Ironstones in both counties are found in the argillaceous formations over and underlying various valuable seams of coal too numerous to mention. The superior quality of Lilleshall Pig owes its celebrity, to some extent, to the excellence of the carboniferous basins which are found under the mountain limestone which overlies this interesting locality. The distinguishing characteristic of the Iron produced here is strength, and resistance to the most tortuous and tensile strains in a manufactured state. This pig Iron is kept in stock in the foundries of the engine shops, in all parts of the world, under the direction of the ablest engineers, and used as a mixture for the metal in any casting, where strength is of paramount importance ; it is likewise invariably used as a mixture at the famous foundries in Staffordshire, where soft and chilled rolls are produced for the Ironmasters in this and all other countries where plates and sheets are rolled. The largest and most generally noted concerns in this county are the Lilleshall Company, and the Coalbrook Dale Company, the chief partner in the former being the Right Hon. Earl Granville ; the latter, whose name and family have been connected with Coalbrook Dale for three or four generations, is Mr. Abraham Darby. The Coalbrook Dale Works were established by the late Mr. Abraham Darby, great great grandfather of the gentleman now at the head of this

When Lord Napier entered the fortress of Magdala pig iron of this brand was found in the foundry. Theodore must have obtained it with difficulty to make his guns.

wonderful and time-honoured establishment. As every-
one knows the name of Coalbrook Dale, from the prince
to the peasant, the merchant to the artisan, as well as
bankers, stockbrokers, and even the more secluded sacer-
dotal element of the various religious hierarchies, from
Pio Nono down to the most humble Scripture reader in
London, all these are familiar with the name of Coalbrook Dale, appreciate the productions of its foundries,
and acknowledge that the make and brand of this eminent old firm is a guarantee for strength and usefulness,
good Iron, good workmanship, and the absence of parsi-
monious trickery, either in metal or workmanship, in
any article they produce.   The Coalbrook Dale Company have five blast furnaces at Lawley, Light Moor,
and Dawley Castle, and one large mill and forge at
the Horse Hay, where all kinds of Iron are made, from
a rail bar to a wire rod.   The Great Dale Works are
situated in a lovely valley; the scenery around is
charmingly beautiful, the rising slopes of the banks of
the Severn, wooded as they are with trees and indigenous shrubs to the water's edge, create an effect upon
the visitor as surprising as it is pleasing : we have seen
nothing, either at the lakes of Cumberland or Geneva,
on the Rhine or in Italy, to supersede this unique love-
liness and natural beauty.   Near to the Dale, on the
opposite side of the Severn, the famous long smoking
pipes are made, at Broseley : these comfortable pipes
are well known in London and elsewhere as Broseleys.
Lower down the Coal Port China Works are fixed
on the banks of the Severn; here Rose's Coal Port
China has been manufactured with success for a

The Coal-
brookdale
Company
is very
nearly as
old as the
Carron
Foundry
in Scot-
land.

century. At the Iron Bridge, named after the bridge which spans the river at this point, is a thriving little business town, supported exclusively by the artisans of the Dale factory, and the numerous Ironworkers in the immediate neighbourhood. A few miles distant are the Madeley Wood Iron Works, the property of the Anstice family, for generations known and justly loved in the neighbourhood; here are three blast furnaces. The Iron, deservedly in request for foundry purposes, always commands a high price, even for Shropshire. Near here, at Madeley, are the Madeley Court Works, the property of Mr. W. O. Foster, M.P., of Stourton Castle : all the Iron made here is sent by the proprietors to Stourbridge and Brockmoor, and manufactured at those famous works into S. C. ♛ Staffordshire bars; the plant here is modern and good, was erected with care, abreast with all modern improvements, regardless of expense, by the late Mr. J. Foster; the Ironstone and coal here are first class. The Old Park Works, and Sterchly, have been for more than a century in the Botfield family. Mr. Boriah succeeded the late Mr. William Botfield, at the death of his uncle. The latter, however, being now dead, the Old Park was worked by a limited company, which was superseded by the present proprietors, and the widow of the late Boriah Botfield, a month since, sold the Sterchly Works, with all leases, and other Iron interests in Shropshire, belonging to the Botfield estate, to the Heybridge Company, a young but enterprising firm, for £50,000, the managing partner of which had formerly held a confidential appointment at Malins-Lee Hall, where

The sainted Fletcher is buried in the churchyard here.

The Madeley Wood Works can be seen from the precious spot that covers Fletcher's remains.

he will now, as principal director, conduct the works established more than a century on the late estate of T. & W. Botfield, one of the oldest Ironmaster families of Shropshire.

THE LILLESHALL COAL AND IRON COMPANY

is on the Sutherland estates, situated an hour's drive from the market-town of Shiffnal, which has a railway station in the centre of it, the furnaces being only a short mile from the Oaken Gates railway station. Mr. T. Horton, the managing director, resides at Prior's Lee Hall, the most easy and agreeable approach to which is *via* Shiffnal with a carriage, through a beautiful country, the chief offices of the company being close to Prior's Lee Hall.

The Lilleshall Coal and Iron Company employ the largest number of men of any firm in Shropshire in the Lilleshall mines and collieries, and at the blast furnaces, and the great engine factory; and for many years, in point of magnitude and importance, have occupied, perhaps, the most important and prominent position of any concern in Shropshire. The firm produces pig Iron, hot and cold blast, famed for its quality throughout the world,[1] and a large quantity of coal for the open market beyond that required to supply their own works; they likewise have large fire-brick works, and one of the most extensive and well-equipped engine shops in the kingdom, supplied with foundries, brass-casting shops,

[1] Lilleshall pig Iron was found in the Emperor Theodore's Cannon Foundry in Magdala in Abyssinia on the advent of the English forces under Napier.

and all kinds of machinery adapted to the manufacture of the largest steam-engines of all kinds. A large business is done here in locomotives for colliery purposes, and for this class of engine the Phœnix Foundry has for a long time rivalled the engine shops of Manchester and the Tyne. Here the well-known patent Dawe's steam compound engines, fitted with Holt's expansion gear, are made, and the Phœnix steam hammer, designed and erected by the company, appears to have been extensively adopted at various Iron-making centres. The success of this great factory, and the general approbation accorded to its productions, may, no doubt, be attributed in some degree to the quality of the Iron, none but Lilleshall pig Iron being used at these foundries, the comparative superiority in the strength of which is admitted on all hands.

In the next place, the tools and machinery of every kind for turning, planing, slotting, drilling, punching, &c., is quite abreast with the Clyde, Manchester and the Tyne ; being near, likewise, to the Sneds hill Iron Works, a ready supply of best plates and bars for the boiler yards and smiths' shops, the best malleable Iron is always at hand, and used in everything made at the factory. The Phœnix engine shops are principally employed on portable and large engines and the heavy machinery used in Iron works ; some of the best blast engines in the United Kingdom have been turned out at this establishment, and if we take into account the extent and perfection of the machinery, the superior quality of the Iron, and the high class of mechanics employed here, the celebrity of the

Phœnix Foundry is but a natural result from the above favourable surroundings. Mr. Lloyd is the chief engineer, whose constant novel erections and designs, combined with the administrative genius of the managing director, have brought the Phœnix Works to the present high position they generally occupy in public estimation. These foundries and engine shops are pleasantly situated on a convenient and healthy spot on the estate; the chief engineer resides near to the factory; the mechanics have comfortable cottages, with gardens, all erected by the company, and the rent charged, we discovered upon enquiry when last we visited Shropshire, was very moderate, Mr. Horton, the managing director, exhibiting more than ordinary interest in the social well-being and comfort of the artizans employed at the Phœnix Foundry.

There is a good market on Saturday evening at Oaken Gates, about a mile distant from these works.

The collieries and Iron mines of the Lilleshall Company are the most extensive in Shropshire, and are kept in constant operation for the supply of coal and Ironstone for the nine blast furnaces of the company. The Iron mines are of the kind called argillaceous, the geological formation being very similar to Staffordshire, coal and Ironstone being found in consecutive seams and strata, one above another. Mr. George Jones is the mining engineer of the company. The company produces upwards of 100,000 tons of pig Iron per annum, of the choicest argillaceous quality, the largest portion being made at the Lodge furnaces by the cold-blast process; the Prior's Lee and Lodge furnaces con-

stantly supply with best pigs the Snedshill Co. and Shelton Bar Iron Co.'s Works, and the makers of the highest class Staffordshire brands of finished Iron. W. Barrows & Sons of Bloomfield, B. B. H. Brand in Staffordshire, consume this best Iron more extensively than any other house in Staffordshire.

There are two distinct blast furnace establishments, the first at Prior's Lee, four in a row, which were erected a few years since on the most modern plan, with every appliance to save labour and to continue the output of the old Lilleshall brand. The blast engines were made by the company, and are a beautiful pair of bright condensers with ample power, and do satisfactory duty without accident. The hot-blast apparatus is perfect, the gas ministering without trouble to the generation of caloric, both for steam boilers and hot-air ovens, the latter being constructed on a new principle, which gives heated blast with increased economy to any temperature required. The richness of the ore facilitates a large make of Iron at each furnace, with a lower temperature of the blast than is generally injected at the furnaces of Middlesboro' or Scotland. In looking over this plant we were struck most by various elements of success in the construction of the works.

First. The admirable selection, *ab initio*, of the site on which the furnaces stand, the coal and mines being delivered from a higher level, which saves labour and renders the large supplies of material more easy, the same advantages apply to the deliveries of the metal by trucks to the railway, the pig-beds being near to the main line, which carries the metal on the rails by the Company's own locomotives on to the main line. The level of the pig-beds gives all that can be desired in gradient to save traction power, and consequently expense. The general arrangements and management are perfect.

Our attention was particularly directed to a patent mine stone-breaker which broke the mine, admirably crushing the largest lumps of limestone at one grip, reducing rock into proper-sized fragments for admixture with the materials for the furnaces. This patent machine,[1] with its attendant steam-engine, is fixed at the proper level for the furnace.

THE LILLESHALL COMPANY'S LODGE FURNACES.

We admire the Lilleshall Company's Prior's Lee furnaces, but we must give an extra mede of praise to the five famous furnaces at the Lodge, which are situated on a lovely spot on the estate. We need not describe them, sufficient to say, in situation, construction, erection, and all machinery they are perfection itself, and if the reader will examine the engraving at the beginning of the Guide, he will be able to trace the perfect lines of the works which dominate the whole, from the splendid pair of bright Lilleshall made blast engines, to the most minute erection in this famed establishment. These works are a picture. We were struck with admiration at the effective simplicity of the machinery which supplies the calcining kilns with oil by an incline

[1] By H. R. Marsden, and was made at the celebrated Soho Foundry, Leeds, on his patent plan so generally in use now.

tramway, worked by steam, and which will be seen at the left hand of the engraving ; the calcining kilns will be seen behind, having the appearance of small duplicate furnaces. These kilns succeed admirably, and with their concomitant feeding apparatus are well worthy of an inspection at the cost of a long journey, the saving of manual labour in this department is marvellous—locomotive engines are employed in all departments, removing the slag right away from the furnaces and take the pigs straight to the railway. Limestone, Ironstone and fuel is supplied to the kilns and furnaces by the same motive power. That, however, which is most to be admired here, is the quality of Lilleshall cold blast Iron turned out at the Lodge Furnaces. In quality it stands alone without a rival, and forms the back-bone of strength in all metal mixtures and every engine shop and foundry of eminence in the United Kingdom. For strength and resistance to torsion strain there is nothing produced like it. This is a fact admitted and acknowledged by our most eminent engineers : where strength is required in a casting used in the most costly machinery Lilleshall cold blast pig is a *sine qua non*. A good chilled roll cannot be made without it, and there can be no doubt that the celebrity of the Phœnix Foundry, and the malleable Iron made at the Snedshill works, and the well known B.B.H. brand in Staffordshire, is in a great measure owing to the large quantity of the Lilleshall Company's cold blast Iron used at these three establishments.

The Kettley Company, one blast furnace only, twenty puddling furnaces, with mills and forges, one of

the oldest concerns in Shropshire, make a very good
Iron; and are known in the market for the quality of
their bars, hoops, and wire rods.   Mr. John Williams,
the present managing partner's father, was always loved
and esteemed by the workpeople and the whole of
Kettley.   Mr. Williams walks in the footsteps of his
father in this respect, and lives out in his life all the
traits of a kind, unostentatious, benevolent, and true
Shropshire gentleman.   The late Gabriel Williams, the
engineer, of whom we shall have something to say hereafter, graduated at these works in the last century.
The Snedshill Bar Iron Company is the largest manufactory of malleable Iron in Shropshire, and for plates,
wire rods, and hoops, has an unrivalled reputation
These works were erected and established by Mr.
Samuel Horton and Mr. William Simms; the former
was brother to the late Mr. John Horton, of Prior's Lee
Hall; the latter learnt his trade under Mr. James
Foster, of Stourbridge, and came down as manager to
one of Mr. Botfield's works, at Sterchley, subsequently
joined the above gentleman, their united efforts
created this extensive and successful establishment,
which has contributed so much to the fame of Shropshire plates and wire rods.   They have thirty-five
puddling furnaces, and eight charcoal fires, with
numerous mills, consuming mostly Lilleshall pig Iron,
the quality of which with Mr. Horton's management
has contributed more to the success and celebrity of
the produce of these works than any other circumstance.   The Snedshill plates are justly famed in the
engine shops of Europe, and will bear testing with any .

in the market. They are certainly equal to any plates made in Staffordshire or Shropshire, and far superior to any except one or two highly famed houses, and we should certainly prefer them to most brands known to us. The Snedshill Works were for many years carried on by Horton, Simms and Bull.

The reader may think we have written enthusiastically in describing the social attributes of the Shropshire Ironmaster : nothing of the kind. We knew them all, except the late Mr. Darby, in business, personally; having often had interviews with Mr. William Botfield, of Malinslee and Decker Hill, of whom it might truly be said that he was always a hospitable gentleman, at the latter, and a thorough business man, with great decision, at the former. Of all the counties in England Shropshire is the most noted for unostentatious hospitality, open-hearted kindness, and straightforwardness; these old worthies were a striking type of the genuine Shropshire character, eschewing the outward garnishments and traditional consequence of the proud Salopians, which are often observable at the fine old county town of Shrewsbury, and no doubt the finest and most perfect type of these old Shropshire Ironmasters was the late and ever lamented Mr. John Horton of Prior's Lee Hall, whose nobleness of nature, kindness and goodheartednes, endeared him to all who knew him.

Last, though by no means least important, in this district, is the Lilleshall Coal and Iron Company, which has nine blast furnaces, extensive engine shops, and foundries, &c., notable for the quantity of Iron produced, the quality when made, and the large number

Mr. George Jones took with him into Staffordshire a profound knowledge of coal mines, which was turned to practical and profitable results.

of men employed in their extensive mines, collieries, and other departments, that we have given a separate notice of this far-famed establishment. In concluding this chapter on Shropshire, it is remarkable to notice that, except at Coalbrook Dale, our friends in Shropshire, as compared with Staffordshire, Glasgow, and the young and rising district of Middlesborough, have always been slow in developing metallurgical industries; the district, up to the present time, being without any of those great hives of industry in the hardware trades (the Dale always excepted) which characterize the districts above referred to. The Shropshire men have shown but little inclination to migrate to other districts, the Baldwin family, of Staffordshire and Worcestershire, the Bagnalls of Gold's Hill, and the Jones's, of Shakerly, being the only families of note who emigrated and established themselves in the Iron and tin trades, in the above districts. The workmen through the whole district are quiet, orderly, intelligent, and industrious, loyal to their masters, strikes or disagreements being rarely heard of; the men in this district have always appreciated the paternal treatment of the masters, and work harmoniously with them; we only recollect one strike, which took place at the Snedshill works, in Horton, Simms and Bull's time, about twenty years since. The principal towns proper in this Salopian Black Country are Madeley, the Iron Bridge, to the south; Dawley and Dawley Green, in the centre; and Kettley and Oakengates, in the north; the latter market on a Saturday evening is a singularly busy and interesting one. Wesleyan Methodism prevails to the largest extent in this district. Madeley, Iron Bridge, Dawley and Oakengates, can boast of many large, handsome chapels, for the use of the followers of the sainted Wesley. Nonconformist places of worship abound in the district, and numerous Independents, Baptists, Primitive Methodists, and a few Brethren are to be found among the sturdy, stalwart, honest, Ironworkers of Shropshire. The township of Wellington is situated at the foot of the Wrekin, from the top of which the smoke and flaring blaze of the puddling blast furnaces may be seen by Londoners with peculiarly impressive effect, after dark at night Persons desiring to see this district may descend from the train at Shiffnal; an hour's drive through a lovely country will bring them on to the Lilleshall estate, indicated by a high chimney on the right, in the rear of Prior's Lee Hall, the residence of Mr. Thomas Horton, the managing partner of this Company. The engraving on next page will give an excellent idea of these magnificent works.

COLLEY & CO.'S SCREWS AND BOXES.

Our friends on the Tyne and the Clyde have been enabled to compete with the Black Country in steam-engines and ordinary castings for Ironworks. There are, however, certain specialities which will always be sought for in Staffordshire, simply because there they produce the best; hence, all projectors and proprietors of new works get their fire-bricks from Stour-

'PRIOR'S LEE FURNACES,' BELONGING TO THE LILLESHALL COMPANY.

bridge, their ...
screws and boxes ...
Country, wh... t
the article can
Messrs. Colley &
wich, are ... ...
good, makers of
have often seen t
works at G... ...
made to excel
sending ... ...
Colley & Co 's r
to cut them ...
selection of ...
pins and boxes,
accident to a s ...
until it is ...
that mills ...
be got from Eng

The testi... ...
like John B... ...
pany, and S... ...
out all our own
Co., whose ... ...
he Greatbri...

bridge, their chilled and soft rolls, the wrought Iron screws and boxes, and lifting jacks, from the Black Country, where the manufacture is understood, and the article can be used without fear of accident. Messrs. Colley & Co., of the Hope Works, West Bromwich, are amongst the oldest, and certainly are very good, makers of lifting jacks, screws, and boxes. We have often seen their large screws at work at our own works at Greatbridge. We believe there are none made to excel them ; we should feel confidence in sending screws for a large plate mill to Russia of Colley & Co.'s make. They not only understand how to cut them, but we know they are particular in the selection of the Iron used in the manufacture of both pins and boxes, and we all know the importance of an accident to a screw or a box. The mill cannot move until it is replaced, and it has frequently happened that mills abroad have remained idle until one could be got from England.

The testimonials published in the *Guide* from houses like John Bagnall & Sons, the New British Iron Company, and Samuel Beale & Co., of Rotherham, bear out all our own sense of justice to a firm like Colley & Co., whose boxes and screws we have used ourselves at the Greatbridge Works.

CHAPTER X.

NORTH STAFFORDSHIRE

(Robert Heath & Sons)

Is thirty-six miles distant from Wolverhampton, and has already played a very important part, during the last few years, in Iron making, and is, without doubt, one of the most rising Iron centres in the country. Messrs. Kinnersley were Bankers, and Iron merchants, at Newcastle-under-Lyne. In that capacity, having advanced money upon some mineral property at Kidsgrove, ultimately purchased the Clough Hall estate there, under the advice of the late Mr. Robert Heath. It was soon after discovered that the same estate overlay rich seams of coal. Mr. Heath opened collieries, and in a few years greatly increased the credit and riches of the bank, by the annual proceeds, strikingly exhibited in the balance sheets of the yearly sales of what Mr. Kinnersley was in the habit of calling his "black diamonds." These collieries, under Robert Heath's management, were a source of great profit, and no doubt the accumulation of wealth stimulated the extraordinary liberality in banking accommodation, which the Kinnersleys readily accorded to the master potters of Stoke, and Hanley, fifty years since ; which, if it did not culminate in the end in large profits to the bank, materially assisted to establish

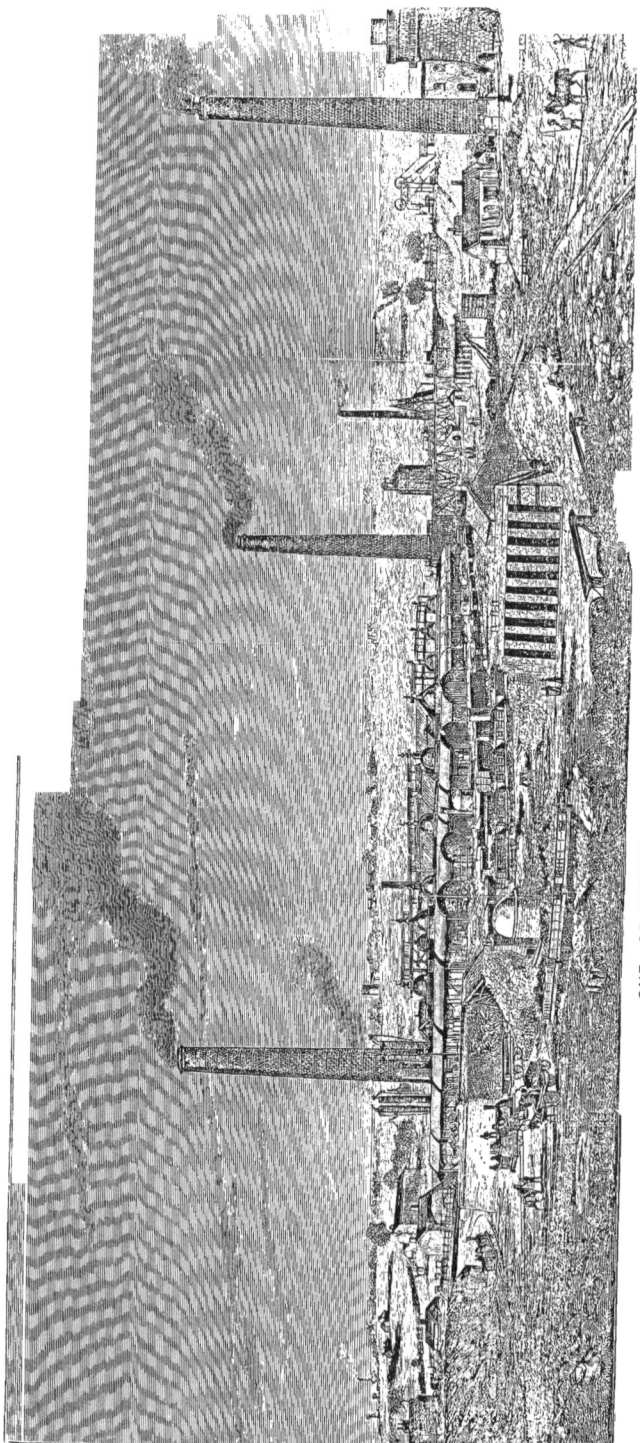

ONE OF ROBERT HEATH & SON'S IRON WORKS, AT NORTON-LE-MOOR, STAFFORDSHIRE

many large china concerns, to extend and consolidate the introduction of the great china and earthenware manufacturers in North Staffordshire, who, by their skill, ingenuity and perseverance, have since raised their productions and created one of the staple export trades of the country. Mr. Heath, sixty years since, commenced making Iron at the new works, at Kidsgrove, which were the first of the kind in Staffordshire; these prospered, and continued to make money for the bank up to the time of his death. Mr. Robert Heath, junior, was brought up here, in the office, with his father: at the old gentleman's death, the present Mr. Heath took to the management, which was equally prosperous during his administration. Twenty years since, Mr. Heath commenced in the Iron trade himself, and has had the most prosperous course of any single-handed Ironmaster in England.

Robert Heath & Son have the most extensive and valuable coal mines in North Staffordshire, and the results of the far-seeing policy of Mr. Heath, during the last ten years, must be marvellous, in the revenues now regularly accruing on the annual balance sheet of their extensive concerns for, during the last ten years, while Ironmasters vacillated, through timidity caused by low prices, Mr. Heath pursued a steady course of purchasing and leasing coal and Ironstone properties, as though he clearly foresaw the good time coming, and should be enabled, in 1872, when these extensive Iron and coal mines were opened for a monster output, to sell both his Iron and coal at just double the price in the Market. This was the case last year.

This firm was not led into heavy contracts, like the

Welsh rail makers, at the beginning of the year; on the contrary, their Iron, coal, and minerals were sold at the current prices, which, no doubt, made by far the most handsome return of any house in Staffordshire, or, perhaps in England (the Earl Dudley excepted). Robert Heath & Son have now, and worked all last year, the Biddulph, and Norton, and Ravensdale Iron Works, which have in all 154 puddling furnaces, with fourteen mills, and eight blast furnaces, at Biddulph, besides their extensive mines and collieries. The works are capable of making 1,600 tons of finished Iron per week, which, at the present average price would amount in value to about £1,257,000 per annum.[1] The works are constructed on modern principles, and the hoops, bars, and plates made here stand high in the market, maintaining their ground with the best leading Staffordshire houses. We have devoted a little more space to this sketch, through a knowledge of the important fact, that whether in mines, or Iron manufacture, this is decidedly the largest concern, employing a greater number of men than any Iron-making firm in Staffordshire.

Mr. John H. Cocksedge is the London Agent. Offices : 90, Cannon Street, E.C.

The Great Shelton Bar Iron Company is situated at Handley, and is the property of the Right Honourable the Earl Granville, Lord Warden of the Cinque Ports, Her Majesty's Minister for Foreign Affairs, &c.

Mr. William Roden, M.P. for Stoke-upon-Trent, is

---

[1] Robert Heath & Sons have at the Norton Works the best mills and machines for making anchors in Iron in England, and are capable of making angles of all sizes of the greatest lengths. For this class of Iron the firm has no rival in this country.

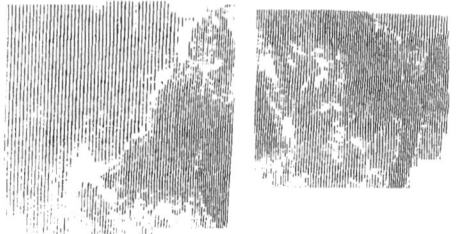

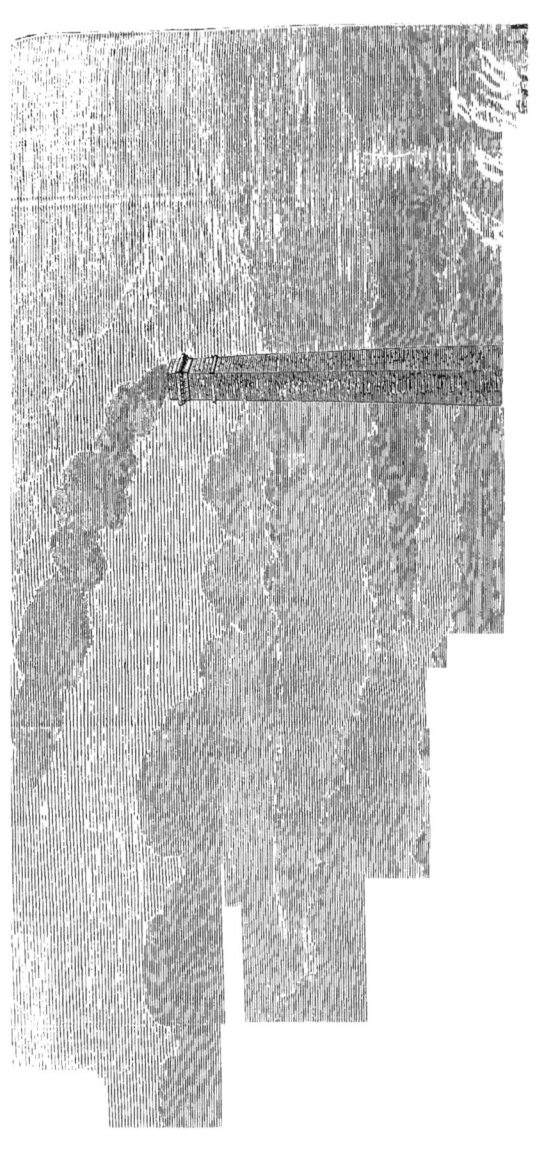

managing partner with his Lordship, in this extensive concern.

There are ninety-four puddling furnaces, seven mills, and eight blast furnaces, with extensive Iron mines and collieries. The works were laid down sixteen years since, under the direction of first-class engineers, and erected with assiduous care, quite regardless of expense. The Shelton Bar may therefore be truly called a model works. The noble Earl had worked the blast furnaces long before the Shelton Bar was established. Most of the pig Iron, during this period, had to be sent by rail to South Staffordshire, which occasionally turned out a very unsatisfactory market. Mr. Frederick Wragge for a long period has acted as his Lordship's sole manager and chief agent in this department. The pig Iron made now is mostly consumed at the Shelton Bar, but as this company have for years been improving the quality and raising the celebrity of their bars, plates, and angles, large quantities of his Lordship's famous Lilleshall brand (the best forge pig Iron in Shropshire) are used at these works as a mixture ; hæmatite pigs from the North, from the most famous makers there, to mix with the other brands above referred to, are all used to produce the splendid plates, for the production of which the Shelton Bar Company is so justly celebrated, at the yards on the Thames, the Clyde, and other ship-building centres in this country. The bars are of a very superior quality, and may fearlessly be compared with the produce of the leading houses in Staffordshire, the Iron being peculiarly suitable for the use of Railway Companies, engine shops, and machinists, in Manchester, and on the

Tyne. The works are adapted to make rounds and squares of large diameters, plates as large and as good as most houses in Staffordshire, the management having, by the addition of Lilleshall, the Barrow Steel Co., and other pig Iron, succeeded in making a quality of malleable Iron, not in the least red short, a beautiful light colour in the fracture, with a rich fibre, and which will stand a tensile strain in a much higher degree than that of many other houses in Staffordshire. The Right Hon. Earl Granville is the chief partner, being associated with W. Roden, Esq., M.P., Thomas Horton, Esq., of Prior's Lee Hall, and other gentlemen. Mr. Roden is the managing partner of the Shelton Bar Company. The Cliff Vale railway station comes very near to the works, or an easy advent is accomplished *viâ* Stoke-upon-Trent. Good cabs are obtained at the Railway Hotel, Stoke-upon-Trent, which is without exception the best hotel in the Potteries, the waiters being very obliging, and the landlady particularly anxious to make her guests comfortable.

The Shelton furnaces are the largest group together in North Staffordshire, containing eight in all. They belong exclusively to the noble Earl. His Lordship has extensive coal and iron mines here. The furnaces are very near to the Shelton Bar Company's works, and minister to a great extent to the supply of pig Iron of the Shelton Bar works, and other Staffordshire houses, who depend to some extent for the supply of the raw material on these furnaces. A very large number of men are employed here. As before stated, Frederick Wragge, Esq., is the manager and chief agent for his Lordship at this blast furnace establishment.

and
good
ving,
. Co.,
ality
beau-
fibre,
igher
shire.
rtner,
homas
emen.
n Bar
very
lshed
t the
hout
aiters
ularly

ether
They
dship
naces
vorks;
f pig
dshire
ply of
large
tated,
agent
t.

SHELTON BAR IRON MPANY'S WORKS AND M LLS

Viewed ley, Staffordshire.

CHAPTER XI.

THE MIDDLESBOROUGH, OR CLEVELAND DISTRICT

HAS surpassed by far all the Iron centres in England and Scotland in the quantity produced last year, being no less than 1,968,972 tons, for 1872, of pig Iron, an increase on 1871 of 84,000 odd tons. The make of this district in 1854 was only 250,000 tons. The first item in these figures, which represents the make of 1872, speaks more conclusively than anything we can write in proof of the marvellous progress Cleveland has made, during the last nineteen years, in the quantity of pig Iron it produces; indeed the rapid progress made, and the extraordinary development of this district, in the interval above referred to, is a marvel to the trade generally, and we should think affords constant surprise to the Cleveland Ironmasters themselves. We must attribute this wonderful and rapid success in Iron making to three causes.

First, cheap Ironstone.

Second, favourable conditions for a supply of coal and coke at a cheap rate.

Third, facilities for shipping the Iron at a moderate freight to all parts of the world. The usual price of Ironstone, delivered at the furnaces, was only about 3s. 6d. per ton. The coal was from the inexhaustible

Original low price of Mine.

coal-fields of Durham and Northumberland, which were near, and cost, before the advance, as little or less than the most favoured districts. The shipping charges were much more favourable than Staffordshire, Shropshire, and Northamptonshire, from their inland position. We believe these triple advantages have assisted Cleveland, more than any other circumstances,

Large deposits of Iron Ore gotten at little expense.

to achieve the gigantic results in quantity referred to above, as the total make. It may be a little surprising to observe the original value of the ore delivered at the furnaces. This, however, is owing to the generous deposits of the metal in horizontal beds, which, in many cases, extend to a great thickness in the mountain ranges, and are frequently run out by an adit or drift from these beds on to the railway, which carries it to the furnaces. The mines generally are not deep, like Staffordshire or Lancashire; they mostly run with a three-foot seam of ore, which is left to form the roof, thus covering the main deposit, which is found six inches lower; this last runs from six to ten, and often fifteen feet thick. The nature of the mine, too, renders the extraction comparatively easy. The Royalties, too, are very low, say from fourpence

Shift system of working the mine by the men.

to ninepence per ton. The men work by shifts, from 6 A.M. to 2 o'clock in the afternoon, and again from 6 in the evening to 2 o'clock the next morning. It is thought and believed that the Cleveland district con-

The Iron Ore will last 70 years.

tains 5,000,000,000,000 of tons of this ore, and that it may supply the district, making reasonable allowance for the probable increase of consumption, for about seventy years. The ore of Cleveland is not so rich in

Iron as some other districts, it resembles very much in appearance and quality the great mine at the Seend Works in Wiltshire, or the Dustan Mine in Northamptonshire, but we have seen no mines in Cleveland anything like equal, in the depth of the deposit, to either of those mentioned, both being of the greatest depth, and most inexpensively extracted of any of the mines in either of these counties.   Having inspected the whole of them, we can speak with confidence on this subject ; one man at one shift, in Cleveland, will load up about six tons per shift of eight hours, and most mines employ two shifts.   The plan of working here is in advance of other districts, and secures the greatest quantity of mine at the least expense.   Three years since, the labour-cost of extraction was from $7\frac{1}{2}d.$ to $10d.$ per ton ; now the former is $1s.$ $4d.$, which raises the price of the ore to the masters, at the furnace mouth, to $8s.$ or $8s.$ $6d.$ per ton.   The difference in the character of the deposit here, renders the extraction easy and inexpensive, compared with the punching and blasting so necessary to disengage the hæmatites of Cumberland from the carboniferous limestone, into the fissures and pockets of which it was originally so mysteriously injected.. The same remarks will apply, though not in the same degree, to the black band of Wales, and all other argillaceous Ironstones, so plentiful in Scotland and Staffordshire, the cost of getting the Flats, Robbins, and Balls, being a charter of no less than $10s.$ to $15s.$ per ton at Bilston at this moment.   These comparisons are intended to exhibit to those unacquainted with the Iron Trade, the great advantage which Cleve-

The two largest deposits we have seen of this kind of ore, either in England or abroad.

Present increased cost of mine.

land masters possess in the nature and position of the Cleveland Ironstone, and that, which is of equal importance, their contiguity to the inexhaustible supplies of coal underlying the Durham and Northumberland districts, which, three or four years since, could be converted into coke, and delivered at the furnace mouth, in Cleveland,

Increased cost of making Iron.

at prices very much lower than coke could be procured at the furnaces in any other district. Things are now changed. The demand upon Northumberland and Durham for 32,000,000 of tons of coal and coke last year, has enabled the coal owners to double the

Coal doubled in price.

price. Therefore, with the present price of coke at 40s. per ton, and Ironstone at 7s. 6d. per ton, the masters will not be able to produce pig Iron now, at less than 86s. per ton, with all the great advantages Cleveland enjoys from fuel, mine, and her seaboard. It is a fact that should be stated, that the Cleveland Ironmasters never use raw coal in the blast furnaces, but invariably coke; on the contrary, Scotland, Wales, Staffordshire, and Shropshire do so to a very large extent. The reasons are obvious, arising out of the different qualities of the fuel used in this and other districts; but space forbids our going further into it in this chapter.

The importance, however, of Iron-making here, will be more fully realised by the reader, when we state that 14,000,000 of tons of coke and coal, Ironstone and limestone, must have been carried to, and consumed by, the blast furnaces, to produce 1,968,972 tons of pig Iron, which again was removed to the puddling furnaces, foundries, and shipping ports. Besides the enormous value of the Iron produced, the freights and charges for moving the mines and the metal must form an aggregate amount astounding to those not acquainted with the trade, and circulate a weekly sum for distribution in these districts greater perhaps than any other known to us. Although the price of Iron has so considerably advanced during the bewildering prosperity of last year, it is correctly stated, and we think the fact may be relied on, that the Cleveland Ironmasters have not made profits commensurate with the doubled price of the metal. The masters in this district, under the influence of the languid demand and low prices of 1867, 1868, and 1869, and other disturbing political elements ahead, entered into large contracts in the closing months of 1871 and the early part of 1872 at low prices, which took them right through 1872 ; and, in many cases, half the make of 1873 was sold, leaving the makers bound to deliveries which precluded them from the advantages of the high prices which ruled in the middle of last year ; and although the Conciliation Board, with Mr. Rupert Kettle's assistance, has worked well with the Iron workers, the vexatious conduct of the miners must have been a prolific source of loss and anxiety to the leading smelters of Cleveland, during the greater part of 1872. The manufacture of malleable Iron has become a great staple trade here, and may almost be said to run, *pari passu*, with pig Iron making. There are in the district upwards of 2,000 puddling furnaces, and corresponding reheating furnaces; the largest works are those of the Consett Company, Weardale Company, and Bolckow and Vaughan's, on the Tees ; besides these,

The Ironmasters make unfortunate sales of pig Iron in 1871, which in some cases went right through 1872.

there are numerous other large manufactories, among which may be mentioned Messrs. Samuelson's, recently converted into a Limited Company, the machinery of which is considered perfect, being erected replete with all modern aids and improvements. Bolckow and Vaughan's, too, are admirably adapted and equipped, with reversing gear to the plate mills, and, on this account, can turn out prodigious quantities of heavy plates in a single *turn*. The rail mills, at all these great works, are well constructed, enabling the makers here to produce rails with as much facility as the oldest districts.

In the beginning of this great Cleveland industry, the puddlers and Ironworkers were principally imported from Staffordshire, and, up to this time, the puddlers watch the prices in the old Black Country, and appeal to this standard as a rule for their own. There are large foundries here, cut nails and the light and heavy casting trades appear to flourish; but the manufacture of tin plates, japannery, and general tin plate goods, has not taken root here yet, and as the Iron produced here is not adapted [1] for tin plates or

---

[1] These remarks must not apply to the Weardale Company; the Weardale Iron is without doubt by far the best made in the Cleveland District, ranking in its market value with the very best Staffordshire makes. This old and highly respectable concern, having the most valuable spathic Iron ores in England, admirably adapted for smelting into Spiegeleisen, and which, combined with a peculiar hæmatite deposit, possessed only by this Company, makes the very highest class of Iron for all purposes, and we believe would make Tray Iron and Black Plate superior to any made even in Staffordshire itself. The Weardale Company have 6 blast furnaces, 60 puddling furnaces, besides reheating furnaces, rolling mills, and four $2\frac{1}{2}$-ton Bessemer converters. With the exception of the Ebbw Vale Company, this is the only firm in England able to make Spiegeleisen from their own spathic ores, which are without doubt the most valuable in England.

trays, probably Cleveland may never rival, in its manufactures, the famous japanners and tin plate workers of Wolverhampton and Birmingham.

The title of the Company is the ' Weardale Iron and Coal Company,' Limited.

They have mines of Iron ore in 'Weardale' which they are working, and smelting at Tow Law near Darlington (where they have four blast furnaces, two of which are out of blast, being under repair and enlargement). These ores, being deposits of spathic carbonate of Iron (more or less) where near the surface, spontaneously decomposed into hydrated per-oxide of Iron, and they yield, as is known to all metallurgical authorities, Iron of a quality which is not surpassed by the best of English or foreign ores. The steel that is made of this Iron is not equalled by any other made in Britain.

They have also two blast furnaces as well as rolling mills at Tudhoe near Ferry Hill. The blast furnaces working partly upon 'Weardale' and Cleveland ores obtained from extensive mines which they hold in the neighbourhood of Guisborough.

The Company have extensive coal mines both at Tow Law and at Tudhoe which are worked for their own consumption, and also for sale, chiefly in the form of coke, for the use of other Iron works.

They make steel rails, tyres, and other important steel and Iron forgings on a large scale. The quality of the Iron and steel they produce, and the high standing of the firm, gives their brand a commanding position in the general market, which generally affords them a good supply of orders even in flat seasons of the trade, when other houses, less fortunate in the above respects, suffer from the paucity of orders. Their brand is TUDHOE ☙. They have an extensive wharf and warehouse on the Thames.—Mr. Robert Troubridge, London Manager. This old firm makes bars and every other kind of Iron.

CHAPTER XII.

BOLCKOW AND VAUGHAN.

WE must not leave Middlesborough without referring at least to one concern in particular which has done so much by liberality and example to establish Middlesborough as the largest Iron-making centre in England. In 1838 Bolckow and Vaughan commenced in Middlesborough to manufacture bars. The difficulty in obtaining suitable pig Iron for their foundries which were first established, drove Mr. Vaughan to think of establishing blast furnaces. In this dilemma, Cleveland for mines was judiciously fixed upon under lease from the trustees of Lady Hewley's Charity of the Eston Royalties. So earnest and determined was Mr. Vaughan to succeed in this new undertaking, that he projected a new line of railway to bridge the difficulty which distance of Eston from the works presented, and although it was only commenced in October, the line was triumphantly opened for traffic on the following 6th of January, 1851. Mr. Vaughan's modest estimate of the mines was 1,000 tons per week. The first year, however, more than three times this amount was turned out. The Eston mines have steadily progressed in their out-put with the Cleveland district, and to-day they yield 2,500 tons per diem. The next step was the blast furnaces at Wilton Park (one recently erected makes five here). Mr. Thomas Vaughan must have the credit of introducing high furnaces of large dimensions into Cleveland, which are now generally adopted in the district. At one furnace at Witton Park, 400 tons of Bessemer Iron is being made per week, the ore being imported from Spain and Whitehaven. Messrs. Bolckow and Vaughan's celebrity, however, is derived more from their extensive manufacture of malleable Iron. They have 110 puddling furnaces, and made in 1852 at these works 56,000 tons of plates, rails, and other kinds of finished Iron. Their machinery for large plates and rails is perfect, and the Middlesborough Works are well adapted to long angles of all sizes. They have eight mills in operation, and ten steam-hammers. Taking Eston, Witton, and Middlesborough together, these works are a credit to Mr. Vaughan, the designer, and an honour to the great Cleveland district, which can so justly boast of some of the most able and scientific Ironmasters in the United Kingdom.

This great company is merged into a Limited Company, and is eminently successful in paying large dividends. The shares to-day are quoted in the ' London Iron Trade Exchange' at 60l. each, 35l. only being paid up. We attribute in some measure, the success of this company, to the judicious expenditure of money in the original erection of the works ; mainly, to Mr. Vaughan's management, and their wonderful mining property. We believe the original proprietors still have the largest interest by far in the concern, and give their able assistance to the management.

CHAPTER XIII.

NORTHAMPTONSHIRE IRON DISTRICT.

NORTHAMPTONSHIRE, during the last fifteen years, has become a very important district for Iron ore; it likewise has, at the present time, twelve blast furnaces, two of which are idle; the Butlin Company being the largest, with four blast furnaces. Northamptonshire last year made 70,500 tons of pig Iron, and raised 1,000,000 tons of Iron ore. The best mines here are the Dustan and the Gayton, the former being almost inexhaustible for quantity, and the latter the best quality in the county; being sound, solid rock, from ten to twenty feet thick, extending under Gayton church and churchyard. Both Dustan and Gayton belong to George Pell, Esq., of Heyford; besides these there are Dean and Chapter, Glendon and Findon, Willenboro', Blisworth, Newbridge, Castle-Dykes, Stonepit Close, East End, Woodford Islip, and King Sutton. The ore raised last year was distributed by rail between South Wales and the Forest of Dean, South Staffordshire (the largest portion), Derbyshire, Yorkshire, and Worcestershire. The ore here is of the oolite kind, and is found only a few feet or yards from the surface, in loose deposits, except at Gayton, where it is a kind of rock band of splendid quality : we have often inspected mines in this district.

Stinson's Malleable Iron Works, at Northampton, make good malleable Iron, which fetches very high prices; and another malleable Iron works is in course of erection. Most of the pig Iron exported from Northamptonshire is consumed in the South Staffordshire district: the Ironmasters' names will be found in the Tabular List. All the coal used here is brought from Derbyshire and Staffordshire, and the coke from Durham. When we carried on blast furnaces in this county our supply for all the furnaces we obtained from the Earl of Dudley's collieries, in Staffordshire, which carried a good burden, and worked well with this ore. The Heyford furnaces will be observed close to the North Western railway, two miles on the London side of Weedon; there are three furnaces here, now worked by Mr. Plevins; the Iron made stands well in the Staffordshire market, which is accessible to the works both by rail and canal. The furnaces are fed with ore by the celebrated Gayton and Duston mines. We must observe that coal has not yet been found in Northamptonshire. An attempt was made thirty or forty years since, by sinking a shaft very near to Northampton a considerable depth. This laudable effort was abandoned, although with our present geological knowledge, the indications of this sinking would not be considered unfavourable to the hypothesis of coal beds below. If coal should be hereafter discovered underlying the Iron deposits here. Northamptonshire may yet become one of the most favoured and successful Iron producing counties in England. We believe coal does exist here.

CHAPTER XIV.

WHITEHAVEN DISTRICT.

THE Whitehaven District is on the borders of West Cumberland, about thirty miles from Barrow, and contains about thirty-six blast furnaces; the Cleator Moor and Workington having six furnaces each, being the most famed for quality in the Frizzingdon district. The Dutton furnace, belonging to Harrison, Ainsley & Co. is the oldest. Most of these proprietors have valuable hæmatite mines at Millom and Frizzingdon, where the best mines in West Cumberland are found; among which we may mention the Salter and Eskatt Park, which is far superior in its revenue to any gold mine. Mr. Thomas Browne, managing director from its commencement.

We have often felt relieved to find our advent accomplished at Whitehaven railway station. This railway is the property of the Barrow Steel Co. The last thirty or forty miles from Carnforth traverses the hæmatite district, being over a monotonous country, which soon convinces the traveller that the traffic in minerals and pigs is of no ordinary kind, and must of necessity require great care and expense to keep the railway in efficient working order. The Millom mines and furnaces on the left, and nearer still, the College of St. Bees, the latter educating, the former

THE PORT OF WHITEHAVEN.

The two Shafts or Chimneys right and left of the Port are connected with the Engines which work
Lord Lonsdale's Pits for miles under the Sea.

being a practical illustration of the development of the power of human intelligence over inanimate matter —are the first real indications to the traveller of an early advent to Whitehaven ; omnibuses meet the trains from London. Mrs. Moat's Globe Hotel is comfortable and well-managed, and, for a visitor, the best in Whitehaven. Private rooms and beds all that could be desired, ' *ménage* ' of the very best quality, served up with quiet and genteel propriety. Although the amiable hostess is not often seen, her watchful management is felt, in the prompt acknowledgment of the sound of the bell, and all other domestic ministra- tions in this homely establishment. Horses and car- riages always ready at the shortest notice, the coach- men are well up in the district, and can tell you the name and fame of all the mines in the distance, as you drive over the fissures and pockets in this mountain limestone formation, into which the precious metal was injected from lower deposits in remote and early ages. How this strong solution of hæmatite ore found its way *ab initio* into the lower strata of the earth's crust must be left for geologists and natural philo- sophers to explain, being unable ourselves to pene- trate farther into the *arcanum in arcano* of nature, with the view of discovering how the ore was con- solidated? How long since the injection? and the various causes which cemented it together *in unum corpus*, at the point of contact of the metal with the walls of limestone, which sustained the solution of Iron before the heat of the crust of the earth upon the metal evaporated the water in solution, leaving the residuum what it really is—Whitehaven hæmatite ore, in a hard rocky state, of superior quality, except the Parks Pocket, belonging to the Steel Company at Barrow. The 'Black Lion' is likewise a very estimable hotel, well managed, its guests looked after with paternal care. This hotel is frequented much by commercial travellers, a good smoking-room, where the wealthy mineral lords of Whitehaven smoke their pipes for an hour in the evening, discuss politics, converse on sensible subjects, invariably eschewing town scandal, sedulously avoiding the shop altogether. This select company are always pleased with the landlord's presence, in this very agreeable smoking-room. Very different is this to the small talk and ignorant twaddle generally the concomitants of suburban places of the same kind about London. The first thing that strikes a stranger at Whitehaven is a lucrative old colliery, the property of the noble Lord Lonsdale, which enriches the noble proprietor with the black diamonds, brought by gate-roads from beneath the nethermost depths of the sea, the workings extending to a considerable distance under the watery element at low tide. Most of the coal here belongs to Lord Lonsdale, who turns out about 250,000 tons a year ; these, like the Earl of Dudley's collieries, return a princely revenue from the Iron consumers of this district. His Lordship's agency has always been liberally and well conducted, particularly during the recent coal famine, on the live and let live principle, and as his Lordship's coal is a ' *sine quâ non* ' for the success of the Iron works, too much credit cannot be given to Lord

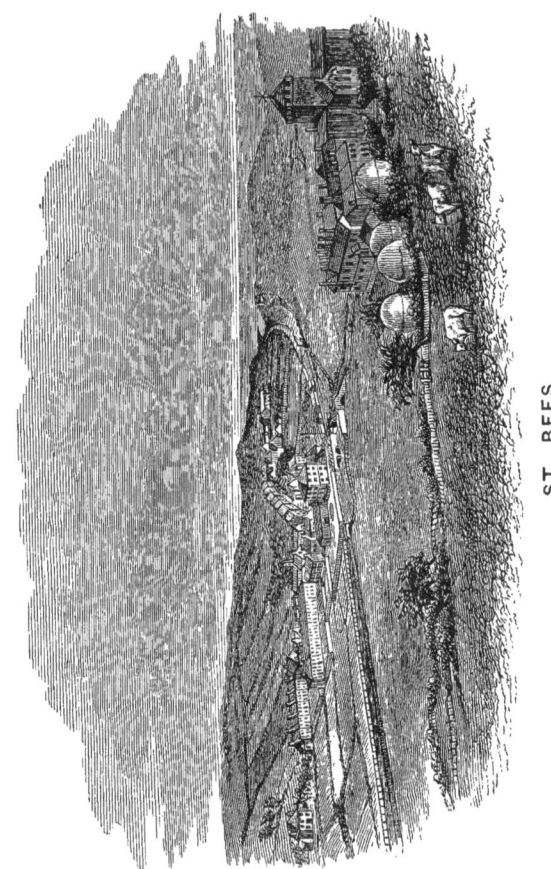

ST. BEES.

COLLEGIATE OLD CHURCH, ST. BEES.

Lonsdale's local administration on this behalf. The present agent is R. Alleyne Robinson, Esq., who is deservedly popular with all classes. The best mines are held by native proprietors, who quietly receive their enormous revenues without ostentatious boasting. Invariably the proprietors understand mining operations, and although regular managers are connected with the mines, the proprietors themselves are to be seen leaving Whitehaven by early morning trains for the various mines, which minister so materially to the wealth and prosperity of Whitehaven. The Cleator Moor, Work-lugton, Harrington, Maryport, and Millom Iron Works, are the most important in this district, all ministered to by hæmatite mines of almost priceless value.

About this time last year we were instructed on behalf of some wealthy clients of ours in the Iron trade to give a fabulous sum for one of these mines, which was rejected, and the proprietors have reason to be thankful that they refused the tempting offer. A list of all the Iron works will be found in the proper place of the 'Guide,' and with regard to these wonderful mines we present all the best of them in a list, and although we must not particularise their relative values, which are well known to us, the consecutive order in which we have placed them may give the reader some idea of their value relatively. Taking these twenty mines as a whole, the value of their output surpasses that of any other district, except Barrow-in-Furness, we are acquainted with,[1] far superseding in revenue the diamond mines of

[1] The Steel company's mines at Barrow, say the Parks Pocket and the Stank, are infinitely more valuable than anything at Frizzington being unrivalled by any mines in any country.

Golconda and the Cape, the silver of Mexico, or the gold of Peru, Australia, and California. There is an Iron shipbuilding yard here, carried on by a company of Whitehaven gentlemen, which is well managed, and turns out good ships of small dimensions. There is a salubrious promenade, the base of which forms a breakwater to the harbour. A stroll from this invigorating spot presents the sea fully to view, Lord Lonsdale's sea-coal pits, and the interesting seaboard of Whitehaven, with three Iron smelting furnaces carried on by the Lonsdale Company.

Further on up the seaboard are the great works of Bain and Patterson, at Harrington, and in succession higher up the seaboard are the Mossbay, the North of England, the West Cumberland, the Workington, the Maryport, Hæmatite Company, and the Solway Company's blast furnaces, all busily engaged in making best Bessemer pig Iron. Lord Lonsdale's pits, the Wellington, and the William, are fixed on each side of this interesting harbour, the gate-roads extending for miles under the sea, whence his Lordship extracts very large quantities of coal, no doubt at a good profit. The noble proprietor of these submarine workings pays a royalty of 4d. per ton to the Crown for coal extracted from beneath the sea. The increased business at this 'little port entails the necessity of new dock accommodation, which is now being actively carried out, and it is thought that the docks already in course of construction will be inadequate to the increasing requirements of this little sea-port.

There are three banks here, the most important

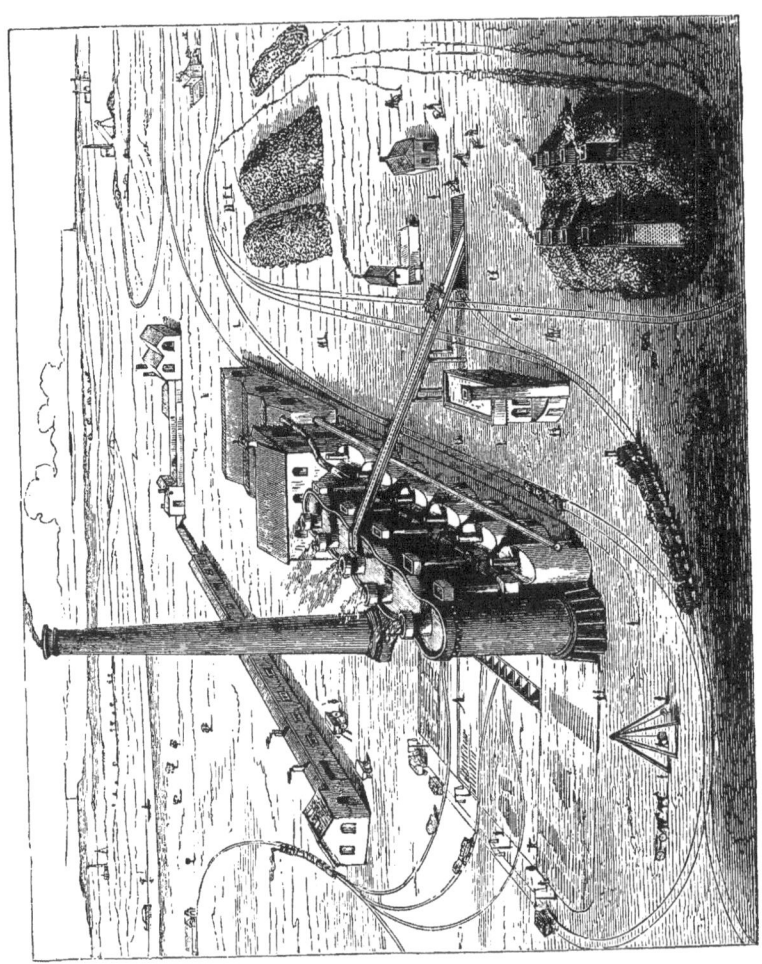

THE PARK FIELD FURNACES, PARK FIELDS, WOLVERHAMPTON.

being the Whitehaven Joint Stock Banking Company, managed by Peter Cameron, Esq., the most able and experienced bank manager in Cumberland. The dividends of this establishment always quadrate with the astuteness of the management.

THE PARKFIELD FURNACES

are situated at Parkfield, near Wolverhampton, being named after the ancient Park Hall Collegiate School, an old Catholic foundation in the immediate vicinity of this estate, distant one mile from Wolverhampton, and about the same from Bilston. These great works were originally erected by J. Underhill, who was subsequently joined by John Bishton. Since the death of Mr. Underhill, they had been worked by other proprietors up to '52, when the property was purchased by Mr. Edward Bagnall Dimmach, in co-partnership with Mr. Henry Martin, Mr. Dimmach's son-in-law.

Enlargements and additions have from time to time been made by the present firm. The furnaces are well served by their own locomotives, on their own roads. a siding into the Stour Valley Railway, and their own Iron roads to the canal wharf. There are five blast furnaces, a pair of splendid beam blast engines capable of driving all the furnaces.

These engines work on the expansive principle.

The arrangements for drawing off the gas here we have always considered the most perfect in Stafford shire. This was one of the first large concerns to

economise fuel in the reapplication of the gases, and
Parkfield furnaces being the largest group in South
Staffordshire, have for years attracted visitors to witness
the successful manner in which Mr. Henry Martin
utilises the gases.   The mineral estate occupies a very
large area, has walls all round encircling the valuable
coal and Ironstone mines which underlie the Parkfield
estate, the property of Henry Marten, Esq., Parkfield
House. The Iron made here is grey forge, and although
some cinder is used, is of good quality.   The quality of
the coal here gives Parkfield Iron a peculiar ' Bell
Ringing ' sound on the ' Breaker,' and as it is an indispensable mixture for most kinds of manufactured Iron,
the large output of these works is always readily sold
in the district.   The management being first class, the
brand is always uniform, no doubt in a greater degree
than any other brand of Iron of this class made in
Staffordshire.   Mr. Edward Bagnall Dimmach formerly
carried on the great Pontypool works in Wales, and
during this period served the honourable office of high
sheriff of the county of Monmouth.   Mr. Dimmach is a
county magistrate both for Monmouth and Staffordshire.

CHAPTER XV.

GLOUCESTERSHIRE, THE FOREST OF DEAN, WILTSHIRE AND HAMPSHIRE.

GLOUCESTERSHIRE has nine blast furnaces, the Cinderford at the Forest of Dean being the most famous, the property of Mr. Henry Crawshay, brother of Mr. R. Crawshay, of Cyfartha, in Wales. The Iron made here, being of brown hæmatite, is valuable for steel making, and in great request for tin plates. Gold Brothers are likewise making very good Iron at Soudley. Oakwood belongs to the Ebbw Vale Company, but is not in blast.

Wiltshire has seven blast furnaces. The Westbury Company make a fair quality of Iron, resembling the Northamptonshire four blast furnaces. Messrs. Malcolm[1] have three furnaces at Seend, with a large deposit of oolite Ironstone; perhaps the largest, most easily worked, and at the least expense, of any that we remember to have inspected; the Iron produced is of the Northamptonshire quality.

Hampshire has only one blast furnace, which is now idle.

Somersetshire has one likewise, at Ashton Vale which makes very good Iron, worked by the Ashton Vale Company.

[1] The proprietary is changed here.

THE HYDE IRON WORKS, ESTABLISHED BY THE FOLEYS.

Messrs. Lee & Bolton are the proprietors of the works represented in the annexed beautiful engraving, which are situated in a lovely valley about four miles from Stourbridge, and are certainly, with their surroundings, the most picturesque of any Ironworks in the United Kingdom. The history of these works will be as interesting to the readers of this chapter as the picture represented by our engraving is charmingly beautiful.

The Hyde Iron Works were erected and for many years carried on by the ancestors of the present noble family of the Foleys—and here we may say this family is noble in nature as well as by patent-title from the sovereign, and has a patent enshrined in the love and affection of the whole hearts of the inhabitants of the district for miles and miles around their territorial domain. As we stated above, the Hyde works were built and worked by their ancestors. At that time Iron-making was in its infancy—this country was far behind Sweden and Russia—we had not then learnt how to make Iron with coal. This was long before Mr. Abraham Darby, Mr. Rennolls, of Ketley, and Dud Dudley utilised coal for Iron-making ; and although at this time nails were forged under the hammer in the neighbourhood of Gornal, Dudley, and Brierley Hill, all the slit nail-rods consumed in the country were imported from Russia, and the price paid for them was 36l. to 40l. per ton, simply because we could not slit them. It is true that Cort had invented rolling in 1792 or 1793, but it was

e works
, which
s from
ndings,
United
as in-
picture
autiful.
r many
t noble
family
om the
ve and
he dis-
omain.
ilt and
ing was
len and
on with
by, Mr.
coal for
s were
ood of
il-rods
Russia,
er ton,
ce that
it was

hire.

Lee & Bolton's Hyde ron Works Worces ersh re.

left for Foley to teach us how to make slit-rods. The difficulty was great; they tried at the Carron, then the first works of the day, but failed; they likewise tried at the Kirkstall Forge and other places: all failed. The appearance of the rod puzzled the best metallurgists of the day; they observed the ragged edges, but were lost when they attempted the *modus operandi* which gave long nail-rods, straight, evidently never cut when cold; and then came the question how it was possible to cut these long rods so thin, hot, without softening the shears. Here England, Scotland, and Wales failed and gave it up; and our ancestors continued to pay 40*l.* per ton for Russian nail-rods.

As we before stated, the Hyde works were thriving under the Foleys. One of the sons, who had been brought up in the works, evinced a great passion for improving the manipulations, particularly of smaller sizes of Iron, and, although the family were wealthy, was continually in the works almost night and day; and as he had acquired a thorough knowledge of music, practised much in his leisure hours on the violin, an instrument in which he excelled, it is said, more than any man in England. Young Foley intimated his desire to visit London, and, after bidding adieu to the family, left with his favourite fiddle—a splendid Sagitarius which belonged to his grandfather: this, we are told, is still preserved among the family heirlooms, and said to be worth 600 guineas. The young gentleman was lost sight of for nearly two years. His friends, although they knew how thoughtful and steady he was, began to feel alarmed for his safety—for it must not be for-

gotten that even London was a very different place
then to what it is now.   Few visitors from the country
were allowed to sleep within the precincts of the city,
and none were allowed to remain beyond a certain
time, always fixed by the authorities.

One evening, about six o'clock, young Foley, care-
worn, copper-coloured, tired and travel-stained, arrived
at the Hyde Works with his beloved fiddle in a green
bag, and a roll of papers carefully wrapped up in cloth
and tied at the ends, resembling a lot of plans of
mines and minerals.   Of course the advent created a
sensation ; and it now turned out that this persevering
young Ironmaster had travelled from Stourbridge to St.
Petersburg, and from St. Petersburg to the Ural mountains, and by the enchanting melody brought out of his
instrument, so fascinated the Muscovite Ironworkers as
to get to see their works—the only foreigner that ever
was permitted to enter a Russian Ironworks which contained a slitting-mill up to that time.   But our readers
will be surprised to hear that Foley laid siege to the
mill no less than two months before he was permitted
to tread the precincts of the cutters, and he might
never have succeeded, but for a singular circumstance,
which opened the portals of Vulcan to him.   Foley's
money was exhausted at St. Petersburg ; any effort to
communicate with England for supplies would reveal
his family connections and perhaps his object ; therefore
he resolved to look to his violin, which he loved so
well, and with this heralded his advent from village
to village, to the manifest delight of the long-skirted
Russians with whom he continually came in contact.

At one place the priest ordered him into prison, but having heard so much of his music, came to hear him, and was so pleased with the melody as to order his release. When he arrived at the Ironworks, he found numbers of huts outside the works, where the men lived. The first day he fared badly, but the next day, Pietri Orloff, the principal man—as it turned out afterwards—at the mill, took him in and gave him a good dinner composed of boiled corn and tallow, which Foley declared these men were fond of. He likewise stated that they eat train oil. He knew French very well, and began to teach two of Orloff's boys French. He always made his way by French, for the Foleys were buying more of this nail-iron from Russia than perhaps any other English house : it therefore would have been imprudent to have revealed his nationality. Foley slept at Orloff's house, teaching the boys French, and fiddling constantly for his own amusement and that of the workmen, but never attempted to go into the works although he saw the precious nail-rods being carried away on mules' backs and loaded waggons drawn by bullocks. After two months, Orloff's two dogs became so fond of Foley that they refused to stay in the works, which were so infested with rats that the absence of these dogs became a matter of serious inconvenience. The largest and the best one, 'Estav,' would go with Orloff into the works, but invariably got back to Foley as quickly as possible to listen to the strains of the music. In this dilemma the manager suggested to Orloff that the French fiddler should be induced to play in the works, and be allowed a certain number of kopecks

according to the quantity of rats destroyed, the name
of the vermin now being legion. Foley fell in with the
offer; the French fiddler and the dogs entered the
works. Foley accepted the reward, had a bed stuck
in the office in one corner of the works, where fiddling
was carried on as usual. The dogs soon destroyed the
rats, and the French fiddler became the greatest
favourite with both managers and men at this famous
slitting mill on the slopes of the Ural Mountains. Foley
next undertook to make them some drawings of the
Notre Dame Church at Paris, for which purpose he
obtained paper and materials; and in this way at night,
with no company in the office but the faithful dogs
'Estav' and 'Petri,' did he make his plans of the Russian
slitting mill. We must now introduce our readers again
to him at the Hyde Works. The object of his absence
soon became known. The cutters were made, the mill
was erected and commenced work; but, alas! it was a
failure, it would not slit the rods. By some accident he
had lost one section of his plan, and Foley became taci-
turn, moody, and disconsolate. Without the slightest
intimation. after a lapse of six months, he absented
himself again, and managed once more, with fear and
trembling, to brave all dangers and fear of suspicion,
manfully embarked on the journey, fiddling his way again
to the same Iron works, where he was received with
open arms by Orloff and the workmen, the dogs not
being the least overjoyed at the second advent of the
Frenchman. He took up his old quarters in the office,
remained at the works twelve months, often worked
the slitting mill himself, made sure of his plans this

time, returned to England, erected the slitting mill, made splendid rods, better cut and of better Iron for nail purposes than the Russians, and what perhaps was at that time most pleasing to him, introduced a new manufacture into the metallurgical industries of his native country. The above facts have been obtained by us from private friends[1] at Stourbridge, who had been known to the father and grandfather of the editor of the 'Guide,' and may be taken as the best history of this marvellous young Ironmaster on record. How Mr. Foley got there and back we cannot say; we know he did not go by a steamer to Ostend or Dover; we know too that he did not travel by rail to London nor by canal; nor did he put his foot into a railway car in Russia, or effect his advent to Moscow other than on foot. We know likewise that he had no steamer to paddle up the Neva and drop him down at St. Petersburg. All this we know, but his privations, sorrow, trouble, and anxiety, with perils by sea and land, we do not know. The Hyde Works, however, established by the ancestors of this noble family, still stand as a memento of the advantages conferred on England by such families as the Foleys and such firms as Messrs. Lee and Bolton, who for many years have been the proprietors, and still make the best slit rods, wire rods, best bars, best best ditto, boiler plates, sheets, singles, doubles, plating bars, and in fact all kinds of Iron of the best quality.

[1] Mr. Rowland Price solicitor, and Mr. Padwick the engineer at the Stourbridge Works.

GEO. BEARD & B.H. EBERHARD, EYRE ST. SHEET IRON MILLS BIRMINGHAM.

THE REGENT'S GROVE AND EYRE STREET IRONWORKS.

The engravings annexed represent the Regent's Grove works, being the largest and most important of this firm, where their famous charcoal sheet Iron of best, best best, and best best best are made. They likewise manufacture corrugating sheets, Russian roofing sheets, Indian sheets, and Canada plates, and indeed all kinds of sheet Iron used for galvanising purposes sugar moulds, kegs, drums, tank-plates, and plates to be tinned. As a manufacturer of sheet Iron Mr. George Beard's practical experience and ability may safely be said to be quite equal to that of any Ironmaster in the United Kingdom. It will be remembered that Ambrose Beard and Sons formerly carried on a successful business at the Regent Ironworks, where the brand of Ambrose Beard and Sons obtained a celebrity in sheets which will always attach to the name of Beard. Mr. George Beard for some years has been in partnership with Mr. Eberhard, who is closely connected by marriage with a wealthy family at Smethwick, well-known in connection with a great metallurgical establishment at that place. Beard and Eberhard at these works continue to manufacture all the specialities in charcoal and other best sheet Iron, for which the old firm of Ambrose Beard and Sons were so noted, and rewarded by prize medals at the Paris Exhibition. The works of Beard and Eberhard are under the constant and exclusive personal management of Mr. George Beard, the former acting manager of the older firm of Ambrose Beard and Sons

REGENT GROVE, IRON WORKS, CAPE, BIRMINGHAM

W. M LL NGTON AND COMPANY'S SUMM RH LL   RON WORKS AND ROLL NG M LLS, T PTON STAFFORDSH RE.

at the Regent Iron Works above referred to, and the commercial department of these well-known works is under the management of Mr. Eberhard. We feel quite safe in saying no firm in Staffordshire is more likely to give satisfaction to high-class buyers than Beard and Eberhard, of the Regent Grove and the Eyre Street works, represented by the woodcuts annexed.

W. Millington and Co., Summer Hill Iron Works, Tipton.

These works were established by Mr. William Millington and his brother, Isaiah Millington, upwards of half a century ago, and are situated very near to the spot where the Tipton old church formerly stood. The works still belong to, and are carried on by Mr. William Millington and his nephew, Mr. Samuel Lees Millington. The latter resides at Wednesbury Oak House, formerly occupied for so many years by the late Phillip Williams, Esq., which will be remembered with interest by most of the old eminent Ironmasters who have, from time to time, visited the late Mr. Phillip Williams, who for the influence he exercised for years over the trade, astuteness in all commercial matters, and sound common sense, had no rival in the Black Country. These works are more useful than ornamental in their external aspect; the works, however, themselves, the steam engines, rolls, and machinery are abreast with the progress of to-day; the boiler plates, and other kinds of Iron made by this firm are known and appreciated in the market. W. Millington and Co.'s boiler plates were honourably mentioned at the Paris Exhibition, with Earl Granville, Shelton Bar, Barrows and Hall (the B.B.H), and the British Iron Companies, and from that day to this the Iron of this firm has maintained a high position in all markets, care being taken here to select good Pig Iron, and Mr. William Millington, who, from the first was a thoroughly practical man, having constantly paid personal attention to the construction of his works, and the quality of his brand of Iron. In our own recollection the works have been enlarged from time to time, and improvements and renovations made, until the completion was obtained as represented in our engraving opposite.

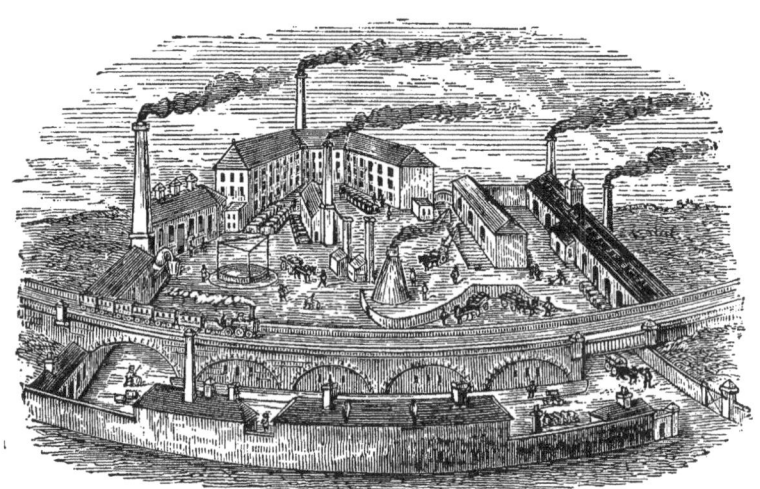

HORSELEY FIELDS CHEMICAL WORKS,
WOLVERHAMPTON.

CHAPTER XVI.

THE HORSELEY FIELDS CHEMICAL WORKS, WOLVERHAMPTON,
PROPRIETORS, MESSRS. WILLIAM BAILEY AND SON.

THESE well-known works were established nearly half
a century ago, having been first opened in 1828,
by Mr. William Bailey, and during this long period
have acquired and retained, through the energy,
industry, and perseverance of the present senior partner
and founder, a world-wide celebrity for the philo
sophical, medical, chemical, and photographic preparations they have produced.

Some of those domestic medicines now to be found
in all countries and in every druggist's shop were first

ıssued to the trade from the above-mentioned works, and a large number of those chemical salts and preparations which, at the beginning of the present century, were merely ' laboratory curiosities,' handled by learned professors with a kind of scientific reverence due to their rarity and costliness, have been, under the influence of gradually improving methods of manufacture at these works, rendered capable of being readily and cheaply produced, so that at last they are supplied by tons instead of ounces, and play no unimportant part in the trade and manufacturing processes of the country. As an example of the former class, we may point to the fluid magnesia, now indissolubly connected with the name of William Bailey; and as an instance of the latter, the curious volatile fluid called bisulphide of carbon may be taken. This was discovered in 1796 by Lampodivus, and has been an object of interest and study to many chemists since that date. Its great solvent powers for India rubber, gutta percha, and resinous gums, its inflammability and general characteristics, fitted it for a variety of purposes in the arts; but its high price proved for a long time a barrier to its commercial use. Liquids costing a sovereign a pound are hardly capable of very extensive application; now, however, bisulphide of carbon can be purchased for less than 6d. a pound, and is consequently manufactured by tons. Shortly after the Great Exhibition of 1851 it was proposed to be used for war purposes by Mr. Wentworth L. Scott,[1] of whose Volcanic Shells and

[1] Now well-known by his researches upon the chemistry of food and sanitary science.

Annihilating Fluid it formed an important constituent. About this time Messrs. William Bailey & Son, the firm who first produced this fluid as an article of commerce, were selling it at about ten or twelve shillings a pound, and since that time their production of it has continually increased, until, as we have said before, it is now dealt in by the ton, and is largely used for the extraction of oils from seeds, wool, cotton-waste, and other substances.

Hyposulphite of soda may be cited as another original *specialité*, for many years peculiar to Horseley Fields Chemical Works, but now largely manufactured throughout Europe. As our readers are doubtless aware, this salt is the sheet-anchor of photographers after nitrate of silver, as, until very recently, it was upon his 'hypo bath' that he depended entirely for the 'fixing' of his solar pictures. In Manchester and elsewhere it is used for what is termed *Antichlor*, by which the excess of chlorine is removed from their bleached goods. This product brings us quite naturally into photography, but space forbids us descanting upon the photographic chemicals manufactured at these works. Suffice that the several preparations associated with the name of that veteran photographer, Mr. Thomas Sutton, are prepared solely at the Horseley Fields Chemical Works.

Whether in relation to medicine, photography, or other applied sciences, the *purity* of chemical preparations is, of course, an all-important point, and it was owing to this quality that a medal was gained by Messrs. William Bailey & Son's preparations at the

International Exhibition of 1862. This firm has for
years been entrusted with the manufacture, on a large
scale, of chemicals for the War Department and Post
Office; and more recently the telegraphic section of
the latter has sought supplies for its batteries from the
Horseley Fields Works.

A special list of telegraphic chemicals informs us that
no less than forty-six acids, salts, and other preparations,
are manufactured for the uses just indicated, and supplied
to Her Majesty's government and the principal railway
companies. Mercurials, an old *specialité* of the firm,
hold a prominent place herein, as also in the Pharmaceutical List proper. Beanes' patent brewing material,
and Bailey's universal finings, as sanctioned by the
Board of Inland Revenue for the use of brewers, are
also among the articles manufactured at these works.
The most important preparation of all, however, in
Messrs. William Bailey & Son's manufactory is, perhaps,
their patent bisulphite of lime, the uses of which, as a
preserver of fresh meat, a regulator of the fermentation
of malt liquor, a restorer of musty casks, and a preventive
of cattle disease, are becoming more and more appreciated in all civilized countries, and must be tolerably
familiar to our readers.

Our hurried notice of Horseley Fields Chemical
Works shall conclude by narrating that the latest
event in its history was the visit of the Burmese Em-
bassy and suite, which took place in November last,
when their excellencies spent a considerable time in
inspecting the various processes and appliances, and
expressed themselves as having been more interested

therein than in anything else they had seen in the Mid land counties.

Being ourselves members of the Pharmaceutical Society of Great Britain from its foundation, and having studied chemistry at one of the principal laboratories in Wolverhampton for two years, during which time we were permitted to visit and improve ourselves in the art of practical chemistry at this famous laboratory, we have much pleasure in giving a brief sketch of this great establishment, which has a world-wide fame, and which has always supplied us with acids and other chemical products employed in ferruginous and metallurgical analysis, purity of the chemicals being a *sine quâ non* for correct determinations.

BLAKE'S STONE-BREAKERS AND ORE-CRUSHERS.

So WELL known are these machines (Blake's Stone-breakers and Ore-crushers), manufactured by H. R. Marsden, the sole proprietor, at the Soho Foundry, Leeds, that it is often the case that journals in the trade, and out of the trade too, in noticing any new improvements of the same, say, ' Of the machines themselves it would be superfluous for us to speak, their merits being so wéll known.' Now we can readily endorse the statement as to their being well known, and also that they are in extensive use all over the world ; but when we consider the numerous uses to which these machines can be put, we are bound to say their merits are not so well known, or rather appreciated, nor áre they in such extensive use as they ought to be, looking at their intrinsic value. It is indeed surprising to many to hear that Marsden's Stone-breakers are used in the remote, and till lately to some extent the unapproach-able, islands of Japan, that Marsden sends out his machines for use on the Ceylon and Cape railways, under the auspices of Her Britannic Majesty's Govern-ment; that Marsden's machines are dragged up the mountainous heights of Bolivia and Peru by mules for

ore crushing. With this view we illustrate, on a small scale, one or two of Mr. Marsden's numerous machines; the first is the fixed machine, with revolving screening apparatus, and fitted with H. R. M.'s new patent cubing jaw for the production of road metal, of a form, as the name of the jaw implies, best suited for road making.

This machine ought certainly to be used in every town; ship and highway district ; when it is remembered how high the price of labour at present is, and that the cost per ton for hand-broken stone is at any rate from 1s. 6d. to 3s. in different districts, for labour alone, and that by these machines the cost is not more than $1\frac{1}{8}d$., or to go to the outside $1\frac{1}{2}d$., per ton, we think it will be readily allowed we do good service by urging this matter upon the notice of all corporations, road trusts, and highway boards. The next machine we illustrate is of the same class, but instead of being fixed it is upon wheels, with horse-shafts, to travel, and can also be so arranged as to be combined with Marsden's Steam Road Roller. Then he has a smaller machine, upon feet, for use when less quantities are required, and when it is desirable to dispense with steam and use only hand power ; in fact these can be worked by hand or steam power as desired ; they are very useful for sampling, or even for doing the same class of work as the larger and more powerful machines. The fourth machine we illustrate shows great economy of space, being an engine and machine combined, for use where space is an object, or where driving-belts and gearing are objectionable. Mr. Marsden has machines in many varieties, all designed for the purpose of breaking or crushing, as needed, the hardest of materials to any required size ; and lastly, for his numerous patents and improvements he has received, at the hands of the best judges at such as the R. A. S. shows, and other exhibitions, upwards of thirty first-class gold and silver medals. Our readers should enquire for themselves,

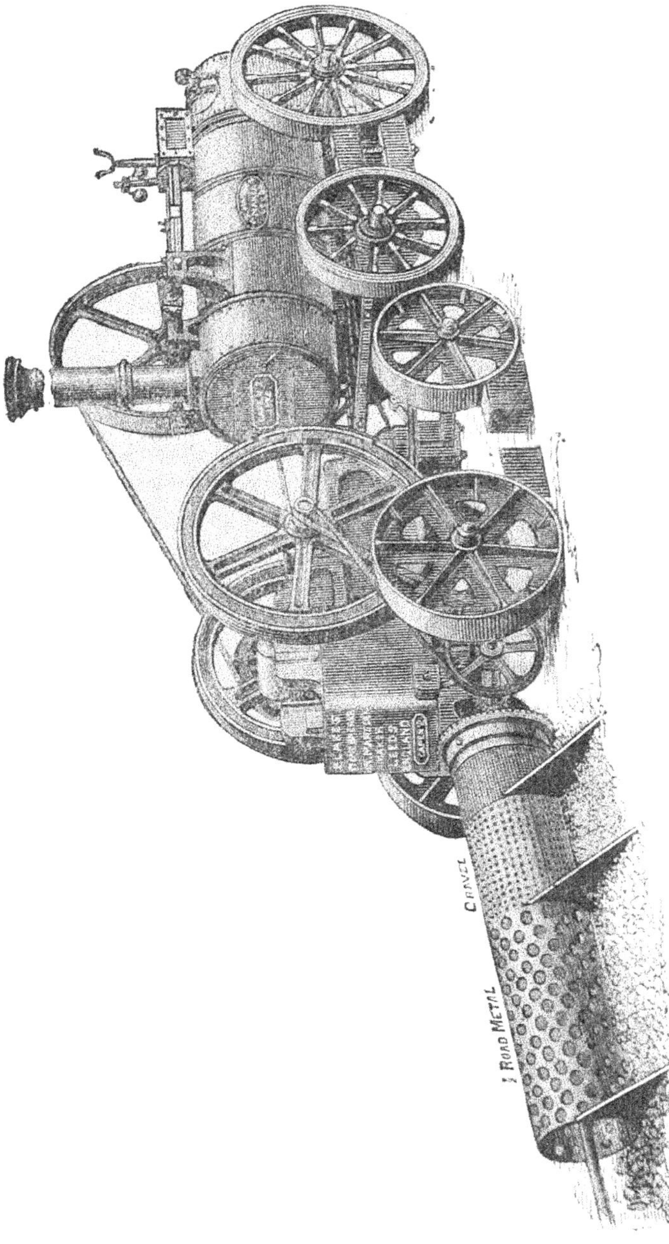

BLAKE'S STONE-BREAKER.

and if their enquiries are satisfactory, adopt whichever machine their peculiar work requires, and they will find that they are able to produce *more* work in *less* time, with *greater* economy, in *smaller* space, and with very much less cost of labour, than heretofore.[1]

---

[1] A short time since we witnessed with pleasure one of these large machines in operation at the noble Earl Granville's Prior's Lee Blast Furnaces, where the monster crusher did its work marvellously well upon the Argillaceous Shropshire Ironstone, so largely consumed at these famous Iron works.

CHAPTER XVII.

NORTH AND SOUTH WALES IRON DISTRICT.

NORTH WALES has 10 blast furnaces, 2 at Brymbo, belonging to the Brymbo Company, 3 at Ffrowd, Sparrow & Poole, 3 at Ruabon, and 2 at Mostyn. The Ffrowd pigs are of superior quality, and fetch a high price in South Staffordshire, being highly approved of by the Staffordshire makers. Brymbo are likewise well known for strength and other desirable qualities. There are 19 blast furnaces using anthracite coal in South Wales; Glamorganshire has 12, Brecknockshire 14, and Monmouthshire 62, which comprise the South Wales group, the Ebbw Vale Company being the largest works in Wales, having 21 blast furnaces in Wales; say at Abersychan, 6; Pontypool, 4; Sirhowy, 5; Ebbw Vale, 3; Victoria, 3; 1 in the Forest of Dean; in all 22; they have likewise 4 large mills and forges, 9 mills for rails and plates, namely, Abersychan and Pentwyn, Victoria, Ebbw Vale and Pontypool, in all 180 puddling furnaces, and 4 Bessemer converting cauldrons, 8 tons each. Their coal and Iron mines are perhaps the best and most extensive in Monmouthshire. Mr. Alderman Curtis, of Manchester, is the chairman of the Company; Mr. Joseph Robinson and Mr. Carter are

the London agents; Mr. Rowbotham, secretary.[1]
This company are expected to make large profits
next year. Next to this Mr. Robert Crawshay, of
Cyfartha, and the Aberdare and the Llynvy Company have the largest number of puddling furnaces
in Monmouthshire. The Dowlais Company have 17
blast furnaces and 161 puddling furnaces at Dowlais;
Fothergill & Hankey have 10 blast furnaces, and 54
puddling furnaces; and the Aberdare Company have
5 blast, and 60 puddling furnaces. These are the
largest Ironmasters in Wales, the bulk of their own
Iron being rolled into rails. While we write the
men are all on strike for wages, all these works having
been closed for a whole month. There are numerous
other large works in South Wales which will be found
in the list in the 'Guide.'

There are several large works in Wales that work on
bars, the Llynvy Vale, Plymouth Forge, Cwmbrian and
Blaenavon; the bulk of the Iron, however, produced
here for years past has been converted into rails;
the largest makers of this article being the Ebbw
Vale Company, the next in this respect is the Dowlais
Company. Without doubt the Welsh Ironmasters as a
body are amongst the most wealthy makers in England.
All old-established concerns, sometimes, in adverse state
of trade, are compelled to work month after month,
almost without profit, with a view of keeping the men
together, and patiently waiting for the advent of better

---

[1] Mr. Grove, the late secretary, has taken an important position, at
the request of the Board, in Wales near to the works, and Mr. Rowbotham has been appointed secretary in his stead.

prices. The interval from the close of 1866 to 1870, was a specimen of the bad times above referred to. A reference to the table following shows that, although the Iron trade generally was so good last year, there was a considerable falling off in the demand for rails for export, which the Welsh masters depend upon, the United States being conspicuous in the diminution of requirements.

Rails Exported in 1872 to the Countries below	Month		Year	
	1871	1872	1871	1872
Railroad Iron of all sorts—	Tons	Tons	Tons	Tons
To Russia	894	7,261	78,367	106,305
,, Sweden	801	1,114	10,918	12,272
,, Germany . . .	648	939	50,287	50,275
,, Holland	255	1,373	14,868	9,026
,, France	6	29	3,653	2,120
,, Spain and Canaries .	1,173	1,883	12,199	11,010
,, Austrian Territories .	16	51	24,260	7,988
,, Egypt	2,400	1,556	16,759	14,472
,, United States . .	37,372	31,686	512,277	472,760
,, Spanish W. India Islands	143	547	3,848	2,315
,, Brazil	1,124	1,578	20,519	20,710
,, Peru	2,337	1,968	29,262	34,874
,, Chili	98	133	11,130	2,845
,, British North America .	216	1,044	61,961	77,248
,, ,, India . .	1,219	1,047	34,523	14,652
,, Australia . . .	964	2,937	14,691	25,091
,, Other Countries .	9,461	10,792	81,675	83,585
Total . .	59,127	65,938	981,197	947,548

CHAPTER XVIII.

SCOTLAND.

WHATEVER branch of trade our good friends beyond the Tweed take up, a combination of perseverance and shrewdness (the most prominent feature in the Scottish character) enables them to run a fair race with, and often rival us in great industries. The ship-yards on the banks of the Clyde, the jute mills at Aberdeen, and the industries of Paisley, all corroborate, in an eminent degree, this view of the question. Iron is another instance of the successful perseverance of Scotchmen.

The time-honoured Carron Foundry too, in the thousand and one articles fashioned here, defies the competition of England, and regularly makes its fair share for the requirements of England and other countries. Since the invention by a Scotchman, of hot-blast smelting, the Scotch Iron trade has been extended in a marvellous degree, and the consumption of No. 1 Scotch Iron, both at home and abroad, has increased in the same ratio, 616,933 tons.[1] There can be no doubt that the quality of certain brands of Scotch pigs are far superior, for foundry purposes, to any other made in this or any foreign country. The superiority must be attributed to the variety of the kinds of Ironstone which

[1] Exported last year.

abound in Scotland, and the admirable adaptability of Scotch coal, under proper manipulation, to the production of No. 1 Iron. The old districts of South Wales, Staffordshire, and Shropshire, make very little No. 1 ; and although Staffordshire has no less [1] than 198 blast furnaces, only 18 are at this moment working on Nos. 1 and 2. Out of these Messrs. Grazebrooks make a very old speciality, which does not compete in the general foundry trade, being cold blast and very high in price. Therefore the number of furnaces in Staffordshire making No. 1 Iron is only 16 : not half the number of the furnaces constantly kept in work by the Scotch makers of the famous Gartsherrie and Eglinton brands ; the reason of this being that the present coal of Staffordshire is not adapted to it : the only fuel left fit for making No. 1 is the thick coal ; this being scarce, makers are unable to compete with Scotland out of their own district : Staffordshire has long since abandoned the melting trade to Scotland. The ordinary melting pig Iron made in Staffordshire at the best, is very inferior to Scotch, and always will be so, the greater part being made from cinder.[2] The young and thriving district of Middlesborough, during the last few years, has manfully kept abreast with Scotland in quantity ; indeed in this respect leaving Scotland in the rear, but in quality it is still behind, and it is questionable whether it will ever be able to produce an article of any-

---

[1] These figures embrace North and South Staffordshire.

[2] H. B. Whitehouse & Sons, of Priorsfield, being rich in thick coal are an exception, and still make the very best melting Iron. Their brand is in high repute, selling to-day at 7l. per ton.

thing like equal value in No. 1 to Langloan, Coltness, and Gartsherrie; never, we believe. The Scotch smelting firms are the most extensive in the world, W. Baird & Co. having no less than 37 blast furnaces, the Coltness Company 12, and the Monkland 9. The largest number in Wales, by one Company (the Ebbw Vale) is 22, and in England[1] 17, the Right Honourable Earl Granville's, chief partner at Lilleshall, where there are 9 furnaces; his Lordship also has 8 blast furnaces at Skelton, in Staffordshire, in all 17 blast furnaces. That which redounds, however, most to the success of the Scotch makers over all others, is their ability to make such a very large proportion of superior Scotch No. 1 pig Iron, commanding the highest price both at home and abroad; the brands to-day of No. 1 being 165s., F.O.B., in the Clyde; the make of Scotland last year was 1,090,000 tons, being a decrease of 70,000 tons on the previous year, owing, we believe, more to the unwillingness of the men to bring the usual quantities of coal to grass than any other circumstance. It has been said that Ironstone is getting scarce. We cannot endorse this hypothesis. Scotland abounds with Ironstone of the very best kinds for foundry Iron; we speak advisedly when we say that the Shott's Company, at Castle Hill, and other leases contiguous thereto, have Ironstone mines which, if opened, would be capable of supplying 50 blast furnaces for 50 years; probably some of the great makers may not be in such a favourable position in respect to mines; on the whole however, Scotland has nothing to fear on this head.

[1] The Great Barrow Steel Company have 16, Bolckow & Vaughan 15.

Coal is another question. The No. 1 Iron of Scotland draws largely, nay much larger per ton of Iron, on the coal mines than any other country, simply because they make most No. 1, which everybody knows consumes proportionately more coal than No. 2, or lower numbers of forge Iron generally made elsewhere ; and if Scotland retrogrades from any other cause than the follies of the colliers, we must attribute it to a deficiency of coal, a good and cheap supply of which is a *sine quâ non* for the profitable manufacture of No. 1 melting pig Iron. We put this as a mere hypothesis, believing as we do that the vast stores of coal in the crust of the earth in Scotland are adequate to the craving requirements of the 127 furnaces now in blast, and, if the colliers will only attend to their work regularly, it can easily be brought to grass, keeping the make of Scotch Iron up to the normal point. In the manufacturing departments Scotland has not kept pace with her smelting furnaces ; she has but 339 puddling furnaces, far below the number of either Staffordshire, Wales, or Middlesborough. There are 14 manufactories, the Mossend Iron Company, at Hollytown, and the Blochain, near Glasgow, being the largest. We have not visited the former ; the latter works, however, we are well acquainted with, and can say, without fear of contradiction, that the works are quite equal to anything of the kind in England, and are capable of rolling large plates and angles in a state of perfection very rarely equalled at any works in Great Britain, the quality of the plates deservedly taking the highest position in the market. Scotland has made little progress at present in the Bessemer process. It is questionable whether she will ever be able to take the lead in this department to the same degree as she has done in melting Iron, the Scotch Ironstone being unfavourable to the production of pig Iron of the sort most desirable for manipulation in the Bessemer pots, owing to the presence of phosphorus in Scotch mineral. We are aware that much has been said and written in favour of Spanish Ironstone, but the fact stares us in the face, that the hæmatite ores of Cumberland and Barrow in Lancashire are the best in the world for this purpose, both in quality and the quantity of the Bessemer Iron they yield ; besides which, these ores are free from phosphorus, which is a *sine quâ non*. There are various reasons in our opinion why Scotland will never be able to compete with Barrow-in-Furness, and the Frizzingdon district of West Cumberland, in the manufacture of Bessemer Iron. Spain, Lancashire, and Cumberland can never give them hæmatite ores to enable them to compete with Lancashire and Cumberland in Bessemer pig, particularly when considerably lower prices rule the market for the metal, which we must expect in due course.

The Steel Company at Barrow-in-Furness have two hæmatite mines capable of supplying all their 16 blast furnaces and to spare. The Park Pocket alone has yielded 365,000 tons annually for the last ten years, and is more promising than ever. The other, called the Stank, is likely to turn out still larger quantities, and Sir William Fairbairn, after carefully analysing this ore, says it makes the best Bessemer Iron in England.

CHAPTER XIX.

THE Carron Foundry is the oldest in Scotland and was established 1760. The manufacture of Iron of all kinds, however, in the whole of Scotland over the succeeding twenty-five years did not exceed 1,500 tons per annum. Cort's invention of puddling and rolling in 1783 and 1784 gave great impetus to the trade. We know from Sir J. Sinclair's statistical account in 1792 that these works consisted of five blast furnaces, made their own firebricks, had a water-engine that worked seven strokes a minute, and raised 3,500 gallons of water at one stroke. The same engine consumed sixteen tons of coal in twenty-four hours. There were likewise three cupolas and air furnaces: the cupolas blown with blast from the blast engine. There were four boring mills to bore guns, cylinders, &c. Anchors, cables, and anvils were made here. There was a forge for making malleable Iron, and a drawing-out forge, and convenience for marking bar Iron with the brand it was intended to bear after its conversion into steel, &c.

The cast Iron hammer and elve weighed one and a half tons.

The original Ironworkers employed here were imported from Russia and Sweden—the great Iron-making countries at this period. The Russian manager put up a slitting-mill, but only to slit the old hammered bars made by the ancient process in use before Cort's Patent. It must be clearly understood that they were unable to slit nail rods here. (Foley must have the honour of introducing this process at the Hyde Works, Stourbridge, still carried on by the well-known firm of Lee & Bolton.)

At this period Russia imported into Scotland 1,000 tons of malleable Iron per annum. The import duty was £3 16s. per ton. The cost of Russian Iron, delivered in Scotland, was £17 per ton, and the best Swedish, Danemora, shipped at Oregund, fetched £24 per ton. The Carron hammered this into charcoal bars, and marked it with the brand for making steel. Other Swedish Iron was imported of less value, as low as £18 10s. per ton. Foreign Iron. advanced very much in price from 1780 to 1791, which stimulated the erection of the Clyde Works, which, we believe, came in succession to the Carron. In 1839 the trade had considerably expanded. The Monkland Iron Company, Dunlop, Wilson & Co., of Dundyvan, the Muirkirk Iron Company, William Dixon, of the Govan Iron Works, were all established in making bar Iron. There were likewise two small forges at Lancefield and the Gartness. These were all the malleable Iron works in Scotland up to 1839. From this period the pig Iron trade in Scotland has increased in an accelerated ratio, but the malleable Iron trade appears now stationary : only 200,000 tons of pigs were consumed last year in this department. The great firm of W. Baird & Co. have not at any time embraced this department of the trade. In concluding this chapter, we congratulate the executive administration of the old time-honoured Carron Company, the founders of which did more to introduce Iron making, steam engines, anchor and steel making in the United Kingdom than any other firm. It is likewise pleasing to observe that the produce of this ancient foundry keeps abreast in the markets of the world with all other competitors.

The brand of Steel Iron made by the Carron Company stood very high with the converters, and after puddling and Cort's rolling were introduced it was a common practice to planish large-sized bars, which could now be made much cheaper by Cort's plan, say from $2\frac{1}{2}$ by $\frac{3}{4}$, and upwards, under the elves of certain ordinary forges known to us in Staffordshire. Spare hammers and anvils were kept in readiness, adapted with a smooth surface for this purpose. After the week's work had been got through the elve was thrown back from the arm so as to modify the blow, and planishing often was carried on until twelve o'clock on a Saturday night, the bars being stamped with a well-known Carron brand and shipped abroad as the genuine article for conversion.

We are glad to say that this practice has been abandoned many years, but that it did exist there is no doubt. We have often seen it practised at Millfields fifty years since, and an Iron-master at the head of a firm of high repute can speak to the same fact, for his father-in-law and Joseph Bladon had the shingling at these works and did the planishing and branding above referred to. Joseph Bladon was the best shingler in Staffordshire, and was father to the managers of this name who have managed at Earl Granville's, Joseph Bull's, Cliffvale, the Bloehain in Scotland, and other eminent works in the North, with so much credit to themselves and satisfaction to their employers. We give this as information in the history of the Trade, believing, as we do, that this very questionable practice is now obsolete.

There are other foundries in Scotland which deserve special notice equally with the old Carron Foundry. Andrew M'Laren and Company, of the Albion Iron Works, Alloa, N.B., and 174 Upper Thames Street,

London, have extensive foundries, and their produce has taken a high position in the London market. The Albion Iron Works are noted for their stoves and ranges, high-class register stoves, and grates of all kinds, of the newest patterns and most useful construction. Ornamental and plain park railings, altar railings balconettes, stair balusters, pilasters, columns and verandahs, including a very large variety of garden chairs, flower stands, tables, and fountains, and all kinds of graceful and elegantly constructed garden work, all kinds of rain water connections, hot water pipes, and gas pipes.

The grace and elegance of the articles, together with the finish and smoothness of the castings turned out at this foundry, places them in a position in this market second to no other establishment in the United Kingdom. Mr. Andrew M'Laren attends to the London business himself in Thames Street, and Mr. John Baker superintends the works in Scotland, the goods being shipped direct from their works in Scotland to their wharf in the Thames. We have often admired their beautiful castings at the great warehouse and show rooms in Thames Street. Their tables are the most elegantly designed and beautifully executed Iron furni ture made at any foundry, elegantly bronzed in the best taste. Their new designs are mostly created here, the very highest class modelling artists being always engaged and at work in Thames Street. The tables above referred to are often seen, indeed generally used to furnish the first-class restaurants of London and the Continent.

THE GLASGOW EXCHANGE.

CHAPTER XX.

SCOTCH IRON WARRANTS.

WHAT is a warrant for Scotch pig Iron? The numerous subscribers to the ' Iron Trade Circular ' [1] have
written to us for an answer to this question perhaps a
hundred times, and although a warrant is simple and
easily explained, our friends from time to time have
difficulty in comprehending how a piece of paper can
represent the value of a thousand tons of pigs, indeed,
how the pigs themselves, the hard metal stacked in

[1] Now called the ' London Iron Trade Exchange.'

Connal's yard, acknowledge their owner when the warrant is signed by Connal and Co., the keepers and guardians of this enormous fold of pigs. Thirty years or more, since, when the Scotch trade had expanded, the stocks had accumulated at the works, the demand fell off, makers wanted buyers, the price of pig Iron was reduced to zero. Merchants and others in Glasgow, took pigs from the makers at these very low prices until the quantity held in this way by merchants and speculators, formed an important item in bulk and value too. These pigs *ab initio* remained on the pig banks of the makers and were bought and sold as a separate commodity, as ' makers' engagements.' The merchants relying on the maker to deliver to bearer, according to the scrip which he held, and which the holder had bought from Mr. A. or Mr. B.

Our Scotch friends are proverbially a little ' canny,' and thought the pigs would be more truly in their own possession at a public wharf; however this may be, we know, as old bankers, that a warrant made out by a public wharfinger for the metal which it represents on the wharf, is as safe for conversion from paper into metal off the said wharf as a Bank of England note is for gold; besides, the Messrs. Connal are kind, obliging, painstaking, gentlemanly and considerate. The holder of any warrant gets it transferred into his own name, or that of his nominee, in a few minutes ; one of the Messrs. Connal, or Mr. Young, being always there in business hours, every facility is rendered, not the slightest indication of ' red tape ' existing in the office. The warrant is presented, and in a few minutes a new name and owner is marked in Connal's register at the small charge of 1*s.* for each hundred tons, and that instant the former owner's legal right vests in the buyer, and the document taken away '*de jure et de facto,*' represents the thousand tons of metal, as much as a note does a thousand sovereigns at the Bank of England. Did we say as much? if you please we will say more so. The warrant is convertible into metal under any combination of circumstances. We cannot say this of the Bank of England, the precious metal is not there to meet *all* the notes out. The pigs, or the metal, are always in Connal's store to meet every warrant out, and any holder may go and examine them as we have done. There they are, stacked and piled in metallic towers. The Messrs. Connal have kept this store for thirty years. The warrant is a more convertible note of the Iron than the bank note is of the gold which it represents. The Glasgow Iron merchants are the most extensive in the world, and as we before stated, they thought it just as well to have the stocks in their own hands : their 'canny' bankers we are sure would commend their prudence in this respect. They appointed the Messrs. Connal their general storekeepers of this Iron, agreeing to pay them so much per ton for unloading, stacking and reshipping, vary-ing from 1*s.* to 1*s.* 6*d.*, at present 1*s.* 6*d.* to 1*s.* 8*d.*, paid by the parties storing the Iron, at so much per month rent for every ton sent, while the Iron remains in their posses-sion, a very satisfactory arrangement, we should think, for all parties. When a large business is being done in and out, we suppose the Messrs. Connal do tolerably well, small as their fees are. When trade falls off, then the revenue of the wharf depends more on the trifling sum paid for rent, but the Messrs. Connal never grumble, but patiently wait for better times and more business.   We should remark here that the freehold of these gigantic yards is the property of Messrs. Connal. Since this store has been established, competitors have started with opposition stores, the railway company being the most formidable.   The willing attention, however, and obliging manners of the Connals leave their stores without a rival.   We think the Railway has about three hundred tons ; formerly the Carron Company held a large stock, but this old stock being cleared off and each of the other makers holding at present but very moderate quantities in stock, Connal's store now contains the only available stocks worth mentioning, and the only pigs which are sold by warrants passing from one merchant to another in no other way than that now in practice, viz., by certificates or warrants, whichever you please to call them, made out by Messrs. Connal the Storekeepers, copied from their register of each lot, which is the legal register of ownership of such lot of Iron being held by the storekeeper for and on account of the registered owner.   The freehold of the wharf being the property of the wharfinger, the Iron is safe against all claims and contingencies, except the legal charge for rent which makes the warrant legal and the Iron safe to the owner against *any contingency*, rent excepted.

It follows that the holder of such a registered warrant has possession of the Iron as securely as if the metal was stacked in the holder's own yard.

A word on the Glasgow Exchange must close our remarks on Scotch Warrants.

The Exchange is a handsome building,[1] abreast in all respects with Glasgow and its Iron-making surroundings. The Iron Market opens at 11 o'clock ; the merchants assemble here punctually, and by a peculiar instinct monopolize any spot they choose in the Grand Area of the Exchange, and a foreign interloper quickly comprehends that he is not wanted in the ' ring.' Numbers of the old firms avoid the active part of the business on the Exchange ; still, they are there to instruct the junior partner or *authorized* clerk who does the business ; there are likewise numerous highly respectable merchants who do their own business. In times of excitement and rapid changes in price the scene in the Iron Circle on the Glasgow Exchange is interesting in the highest degree.

At 11 o'clock all are punctually there, ' bulls ' and ' bears ' rush to the spot, the tallest leaning their heads over from the outer circle, attentively listening to the conversation and offers of the inner circle of the group.

The reader must observe that the group consists entirely of Iron merchants ; probably ten or fifteen minutes elapse without business. The greatest attention is directed by the outer circles of the group, now to the great ' bulls ' in the centre of the ' ring ; ' the throng increases, outsiders feel more interested at a distance : this Iron group becomes now the observed of all observers.

The merchants now appear more intensely anxious,

[1] See engraving on page 169.

standing on tiptoe, eyes and ears exercised to the utmost, as though something very important was trans-piring.   In an instant the physiognomy of the group changes, the anxiety to see and hear ceases, the group partially breaks up, the leading men appear to take breath and indulge in conversation with those around them, and everybody tells his neighbour pigs are up 2s. per ton, 2,000 tons having been sold for cash at this advance, and numerous buyers at the same figure. The first sale fixes the price for the moment.   After a little pause the same throng fix themselves perhaps on ano-ther spot, a repetition of the former scene ensues, tele-grams are seen handed about the Exchange, probably an advance of another 1s. is established; the morn-ing market closes a shade easier. Business is suspended in Iron until two o'clock, when this extraordinary group re-assemble and, in exciting times, an advance of from 3s. to 5s. per ton, in a day, has often been established.

MESSRS. HAYWARD TYLER AND CO.'S WORKS, FOUNDRIES AND FITTING SHOPS IN WHITECROSS STREET, LONDON AND LUTON, BEDS.

The business of Messrs. Hayward Tyler & Co., of Whitecross Street, London, and Luton, Beds, may rank among the oldest treated of in these pages. It appears to have been established in 1815 by William Russell, a pupil of the celebrated Joseph Bramah, and for many years was conducted on a scale not very extensive in Clerkenwell, London. The branches of engineering carried on were those to which Mr. Bramah had given so much attention, and of which he may be almost named the founder, viz., hydraulic presses and machinery for making soda water. In 1827, Mr. Russell's hands were strengthened by the able co-operation of Mr. John Briggs, who still continues his valued and efficient assistance to the present proprietors of the firm. In 1835, on Mr. Russell's death, the business was taken by Mr. Hayward Tyler, who carried it on until his death in 1855, when it was purchased from his widow by Mr. Robert Luke Howard. This gentleman and his brother Mr. Eliot Howard are the present heads of the firm.

The extension of this business has not been on the same very rapid scale as that of some much younger concerns, but the motto of the firm has always been 'slow and sure,' and they have aimed at excellence in every branch which they undertook, rather than at those sudden increases which are often incompatible with the same attention to the quality of the output. It is mainly under the present proprietors that the business has gradually assumed its present dimensions— and until lately the whole of their manufacturing has been carried on in Upper Whitecross Street, London, a locality where few visitors would at first sight imagine what a busy hive of industry was to be found behind the very unassuming exterior. Here may be seen the very utmost economy of space compatible with the convenience of work, and the most intelligent application of modern science and the principle of subdivision of labour to these branches in which the firm have long held a leading position, namely, all the higher classes of brass founders' and coppersmiths' work and hydraulic engineering. Almost every description of pumping machinery is here made, and its strength and finish challenge comparison with any other factory in the world.

Most interesting too is the workmanship of the various classes of machinery for soda water making, in which difficult branch of engineering the firm have for almost half a century had undoubted pre-eminence, and also hydraulic machinery of which they may be said to be the oldest makers, their business dating back to the great Bramah himself and possessing an incredibly varied stock of patterns, amongst others, the original ones of Bramah, which they bought when he died.

The trade, however, which of late years has brought them most before the general public, is their manufacture of Direct-Acting Steam Pumps. Their 'Universal' Steam Pump possesses a great advantage over others in its simplicity; the Engine portion having only two moving parts, and dispensing entirely with tappet-valves and other contrivances so liable to get out of order, as well as with fly-wheels and wasteful gearing.

They have been applied in great numbers and with the most satisfactory results to all classes of deep-mine pumping—sometimes worked by steam, sometimes by compressed air, sometimes slung by chains from the surface and lowered as the water was reduced; at other times placed in 'slants.' The accounts of their performances would be difficult to believe, if they were not confirmed by unimpeachable testimony, such as their continuing to work in repeated instances after they had been 'drowned out' by a sudden rise of water, and actually working themselves high and dry again. It is sufficient to say that, although other makers were well represented at the Vienna Exhibition, these were the only direct-acting Steam Pumps in the English department which received the grand prize 'Medal for Progress,' and the leading engineering papers concur in their testimony of their superiority; thus, on August 1, 1873, 'The Engineer' speaks : 'Although there are a variety of direct-acting Steam Pumps exhibited, none that we have seen work so quietly as those of Messrs. Hayward, Tyler & Co.'; and 'Engineering,' July 11, 1873, says : 'The "Universal" Pump can certainly claim to be the simplest machine of its kind in the Exhibition.'

Messrs. Hayward, Tyler & Co. have lately found it necessary to have increased space for their manufac turing operations, and last year erected a factory at Luton, Bedfordshire, conveniently situated on the high roads both to the north of England and the Black Country, and here they have already a large number of men at work on their various manufactures.

CARPENTER AND TILDESLEY, LOCK MANUFACTURERS, SUMMERFORD WORKS, WILLENHALL.

As Messrs. Chubbs' is the representative factory in the lock trade in what is known as the 'levered' department, so is that of Messrs. Carpenter & Tildesley in what is known as the 'warded' department, of this important industry. The Summerford Works, situate at Willenhall, midway between Wolverhampton and Walsall, were established by the late Mr. James Carpenter, whose business as a lock manufacturer in the town dates from the year 1795. Mr. Carpenter was the first to introduce rim iron into the construction of locks, but his name is better known as the inventor and patentee of the perpendicular motion in the working of lock bolts, the use of which has now become almost universal. In the year 1839 Mr. Carpenter took into partnership his son-in-law, Mr. James Tildesley, the present sole representative of the firm, who personally superintends the entire operations of the establishment.

The Summerford Works give employment to 150 pairs of hands, besides 'out-workers,' and the average production of locks of the rim, dead, drawback, and mortice descriptions, varying in size from 5 in. to

12 in., and in price from 10s. to 100s. per dozen, is something like 250 dozens per week. Of these the greater proportion are exported to the colonial and other foreign markets, where the name of 'Carpenter' in connection with the lock trade has long been familiar as a household word. A noticeable feature of production is the 'double-handed' lock, the invention of Mr. Tildesley's son, and which Mr. Tildesley has secured by patent. The principle of this lock is, that it is equally adapted to doors opening to the right or to the left hand, and both the friction of working is reduced, and the shape of the latch bolt is free from the objectionable sharp angles of that in the ordinary lock. The 'double-handed' lock is made both in rim and mortice, and it already commands a very large sale, both in the home and export markets. Curry-combs to the number of 10,000 per week are also made at the Summerford Works, principally for the United States, continental, and home markets, and a large wood turnery is included in the establishment, door-knobs and curry-comb handles being the leading features of production. Rewards of merit for locks, lock furniture, and curry-combs have been awarded to Messrs. Carpenter & Tildesley at the various international exhibitions where examples of their produce have been displayed.

Correct Particulars of the largest Iron Works in Staffordshire.

Robert Heath & Sons being the largest Iron Masters in Staffordshire, we give detailed particulars below of their Biddulph Valley Coal and Iron Works, the Norton Coal and Iron Works, the Ravensdale (Old) Iron Works, and the Ravensdale (New) Iron Works, including particulars of all the Coal and Iron mines of this important firm, who make bars, hoops, sheets, angle iron, small rounds and squares, and all other kinds of Iron on a more extensive scale than any other house in Staffordshire; their best qualities of Iron being equal to the other leading Staffordshire houses.

BIDDULPH VALLEY COAL AND IRON WORKS

situated at Black Bull, 3 miles from Tunstall, and 7 miles from Stoke.

Coal Pits	7	
Iron Stone Pits	7	
Blast Furnaces	4	(All working)
Puddling and Ball Furnaces . . .	43	
Plate Mill 22 inch, with . . .	3	Heating Furnaces
Bar Mill 18 ,, ,, . . .	2	,,
Bar Mill 10 ,, ,, . . .	2	

1 Patent Three High Mill, capable of Rolling
  20 tons of Sheets in one Turn.

NORTON COAL AND IRON WORKS

situated at Norton, 1½ miles from Burslem 3½ miles from Stoke.

Coal Pits	9	
Coal and Iron Stone Pits . . .	5	
Blast Furnaces	4	(All working)
Puddling and Ball Furnaces . . .	44	
Plate Mill 22 inch, with .	3	Heating Furnaces
Angle and Iron Mill 20 inch, with	2	,,
18 inch, with .	2	

RAVENSDALE (OLD) IRON WORKS

half a mile from Tunstall.

Puddling Furnaces . .	29	
3 Hoop Mills 9 inch, with .	2	Heating Furnaces
1 Hoop and Bar Mill 12 inch .	2	,,
1 Guide Mill 8 inch .	2	

RAVENSDALE (NEW) IRON WORKS

half a mile from Tunstall.

Puddling Furnaces 28
Dank's Furnaces (in course of erection) 9
Plate Mill 22 inch, with Hard Rolls, 3 High Louth's Patent,
  3 Heating Furnaces.
  2 Plate Mills 20 inch, with 3 Heating Furnaces to each.

SUMMARY.

Coal and Iron Stone Pits (now in full work) .	28
Blast Furnaces	8
Puddling and Ball Furnaces	144
Dank's Furnaces (erecting)	9
Mills	14
Heating Furnaces	33

THE PROCESS OF MAKING IRON AT THE BOWLING IRON WORKS.

At the meeting of the British Association at Bradford, Mr. Carbutt read before the Mechanical Science section the following paper on 'The History, Progress, and Description of the Bowling Iron Works,' by Mr. Joseph Wilcock, chief engineer.

There are several indications in the Bradford district that Iron was manufactured here at a remote period of antiquity. It is believed that the Romans both got and worked ironstone in the neighbourhood. Dr. Richardson, the eminent botanist, writing to Herne nearly 200 years ago, stated that Iron was made in the neighbourhood of Bierley, two or three miles from Bradford, in the time of the Romans, as upon a heap of cinders being removed to repair the highway there, he had discovered a quantity of copper Roman coins. The ironstone cropped out in several places, and in many others it lay very near the surface, so that by making 'bell-pits' there would be no difficulty in getting the ironstone. Within a few miles of Bradford there are at work the old-established and still flourishing works of Kirkstall Forge, which claim to have been the first establishment to use rolls for slitting Iron into nail rods, this process having been carried on there so far back as the year 1594. Thus Bradford and the district may claim to have made Roman implements of warfare, and most probably Saxon, Norman, and old English ones likewise. In fact, this department was carried on up to a very recent period, when the Bowling and Low Moor Works manufactured cast-iron guns and mortars. At or about 1784, James Watt was completing his invention of a rotary motion steam engine, the introduction of which was only required to inaugurate a new era in the history of the Iron trade. It was about this time that the Bowling Iron Works were commenced, the first furnace being blown in the year 1788. Even before that date, however, we have records of some part of the works being in existence, and doing a limited trade in foundry and smith work. But as works for the smelting of ores, they date from the year 1788, three years in advance of the sister works at Low Moor. This was the beginning of the trade of the best Yorkshire Irons, now so famous for their qualities through the entire civilised world.

The Bowling Iron Works may properly be considered, therefore, as the pioneer of the great prosperity which has rendered Bradford famous amongst the commercial marts of the world. The population of the borough when the Bowling works were started would only be about 10,000, as thirteen years later, in 1801, it was not more than 13,264, whereas the present population is over 150,000. The establishing of works of this kind, at which employment for a considerable number of men would be insured, must at that period have been regarded as an event of much importance. John Sturges, of Sandal, Wakefield, an Ironmaster of repute, was the first to broach the establishing of Iron works on ground where they now stand, and to his knowledge of the necessary minerals to produce a superior Iron is to be attributed the choice of the situation.

The engine originally erected for blowing purposes was burnt down a few years after it had been at work, and was replaced by the one called the 'Old Blast Engine,' now existing. This was considered to be a great improvement upon the first one, as the valve gear was made self-acting. Below the engine, and constructed in massive masonry work, was made the air chamber for equalising the pressure of the blast. A bar mill and a plate mill were started soon afterwards, and were also driven by a steam engine, a considerable portion of which was constructed on the spot. We find it stated in 'Smiles' Lives of Boulton and Watt' that notice was given to the Bowling Iron Works, near Bradford, of proceedings against the company for the recovery of dues. On this the Bowling Company offered to treat, and young Watt went down to Leeds for the purpose of meeting the representatives of the Bowling Company on the subject. On February 24, 1796, he wrote his friend Matthew Robinson Boulton, as follows:—'Enclosed you have a copy of the treaty of peace, not amity, concluded at Leeds on Saturday last between me, Minister Plenipotentiary to your Highness on the one part, and the Bowling Pirates in person on the other part. I hope you will ratify the terms as you will see they are founded entirely upon the principle of indemnity for the past and security for the future.' On referring to the private ledger of these works of that date, we found that the treaty of peace referred to was purchased at the price of 1,640l.

The sub-stratum around Bowling is part of the most extensive and valuable coalfield in England, stretching from Derby or Nottingham to this district, a distance of sixty miles, and ranging about eight miles broad. The seam of coal called the ' better bed,' which is one of the valuable elements necessary for the production of the best quality of Iron, is seated upon a peculiar hard silicious sandstone termed 'galliard,' immediately above the black bed coal, and resting upon it is an argillaceous stratum of the mean thickness of two yards, in which lies embedded in irregular layers the valuable ironstone of this district. The stone wears a dark brown appearance, and yields about thirty-two per cent. of Iron. Both coals are caking coals and moderately hard. The ash of the black bed coal is of a dark purple gold colour, similar to roasted pyrites. This coal contains a very large percentage of pyrites in a state of intimate mixture in the coal, so that it cannot be seen; the ash fuses readily, is slightly alkaline (due to lime), and containing sulphide of Iron and a very large quantity of oxide.

The works comprise six cold blast furnaces, from which about 360 tons of pig Iron are run per week, five refineries, twenty-one puddling furnaces, forty heating furnaces, an extensive forge, a tyre mill for rolling steel and Iron weldless tyres, one guide mill, one bar mill, with 15 in. rolls, and two plate mills. A third new plate mill is nearly completed. The powerful reversing engines to give motion to this mill are on the principle introduced by Mr. John Ramsbottom, late of Crewe works, and when the mill is completed plates can be rolled of the largest superficial area ever yet attempted.

There is also an extensive steel works for making crucible steel, having about 100 pot furnaces, and which is now in process of extension and improvement by the erection of new furnaces on the Siemens and Siemens-Martin principle, to be worked by Siemens' regenerative gas furnaces. The engineering works comprise foundry, smithy, boiler-fitting, millwright, wheelwright, and fitting shops.

The Bowling Company itself supplies almost all the coal and ironstone which it consumes, its collieries extending five or six miles in various directions, and the main pits being connected together and with the Iron works by tramways worked with wire ropes. The total length of these tramways is 21 miles, the number of pits 42, and the number of hands employed in them is more than 2,000. To work the pits 61 steam engines are required, having cylinders varying from 7 to 70 in. in diameter, and to supply them with steam 81 steam boilers are required of from 10 to 50 horse power each. In the Iron Works are three blast engines, with blowing cylinders, varying from 76 to 84 in. in diameter, and fourteen engines of from 20 to 60 horse power, to give motion to the various machines, besides numerous small engines driving separate machines and pumping water for the boilers. The number of steam hammers is thirteen, and helve hammers two. The supply of steam is maintained by thirty-three boilers of from 20 to 50 horse power each. The number of hands employed at the Ironworks is upwards of 1,000, thus making a total of upwards of 3,000.

The yield per cent. on the raw ore is 32 per cent. iron, and on the calcined ore 42 per cent. iron. The following are the relative quantities of minerals for producing 1 ton of Bowling pig Iron :—Raw ore, 3 tons 3 cwt. 3 qrs. 27 lb.; calcined ore, 2 tons 7 cwt. 1 qr. 26 lb.; limestone, 18 cwt. 2 qrs. 12lb.; coke, 2 tons 5 cwt. 0 qrs. 9 lb. The quantity of pig Iron used to produce 1 ton of bar Iron (finished) is 1 ton 12 cwt. 1 qr. 25lb. The limestone is obtained from Skipton, and is called locally ' Skipton old rock.' The sulphur in all the samples varies only very slightly, and may in fact be considered identical, the difference in the results not being more than those due to the errors of experiment. The phosphorus in all the samples exists in precisely the same quantity, the whole of this element present in the ore combining with the Iron.

I have made a sketch of the original blast furnace at Bowling, now in existence and working to within two or three weeks, presuming it might be interesting to some of the members of the Association. I have been told by some of the oldest inhabitants of Bowling, that there was only one tuyere at first, but two have now been used for many years, the nozzles being $2\frac{3}{4}$ in. in diameter, and the pressure of blast supplied to this and the other furnaces 32 oz.

The Iron for plates and bars is taken direct to the refineries or oxidising hearths. The metal is placed upon the hearth, covered with coke, and a blast is forced over the surface. Two tons of refined or plate metal are produced from each charge, which is run into moulds cooled by water, the refined metal being about 2 in. thick, and 12 ft. long by 4 ft. broad. From the refineries the plate, or refined metal, is taken to the puddling furnaces for conversion into malleable Iron in the usual manner, by charges of about 3 cwt. at a time, and each puddling furnace is charged ten times a day. The quality of the Iron necessitates more attention from the puddler than the commoner classes of Iron, and to insure the extra attention and a uniform quality a premium is given to the puddlers who have produced the best specimens during a turn. The puddled Iron is taken under the steam hammer to knock out the slag and impurities, and is made into what are called ' stampings ' and ' nobblins.' The stampings are broken into several pieces under fall-hammers, piled, heated, taken under a steam hammer, and made into blooms, or billets, in which state they are taken to the bar or guide mill, re-heated, and rolled into round or square bars, angle irons, rods, or such other shapes as may be required. The nobblins are piled, heated, taken under the steam hammer, and made into blooms, or slabs, of various sizes, and afterwards to the plate-mill, where they are re-heated and rolled into plates.

From stampings are made the Bowling Iron weldless tyres. A hole about 5 in. in diameter is punched through the centre of the bloom, forming it into a ring of Iron. The ring thus made is hooked on the beck of an anvil, and is hammered with a suitably shaped hammer head to raise up the flange, the ring being constantly rotated on the beck between the blows of the hammer, so that all parts may be evenly worked. At the end of this process the ring begins to have some resemblance to a tyre, and is then rolled out.

The steel works were erected in the year 1866, and the steel manufactured is crucible steel, produced in the ordinary manner in furnaces heated by coke. The Iron used is scrap from Bowling plates, and its conversion into steel is effected by the addition of suitable quantities of carbon, chiefly introduced by spiegeleisen, and also by a mixture of steel scrap. Of the steel produced, a part is used for making tyres from ingots, in a similar manner to Iron tyres and general forgings; and a considerable portion is used for making castings of all descriptions, where strength, with lightness, is the desideratum. Arrangements are now being made, and are partly completed, for applying Siemens gas process for melting the crucible steel in suitable furnaces; and a Siemens-Martin furnace is also in course of erection for the conversion of pig Iron into steel; the furnace will produce four tons of steel at one operation.

The engineering is done in one extensive range of buildings, where the whole of the work and new plant required to keep the collieries and works described in repair are made. This department is also devoted to the construction of engines, &c., for the market. In the model-room—one of the finest in the country—is a model from which the first wheel was cast for Blenkinsop's locomotive. The boiler shop is now being extended so as to be capable of producing from two to three boilers per week. The foundry has been recently rebuilt upon the old site.

The distinguished qualities of the Bowling Iron are hardness with great pliability, homogenity and uniformity of texture, capability of withstanding the action of fire, and receiving a brilliant polish, it being used extensively in the Sheffield trades on account of the last-named virtue. Works established in the infancy of the Iron trade and producing a superior quality of metal—quality being always preferred to quantity whenever the alternative presents itself—must naturally be disposed to conservatism. Besides repeated experiences have proved the necessity of keeping to the original mode of working with the minerals and Iron. It is rarely known to what purpose or tests the Iron may be put to on leaving the premises, but it is known that it will have to withstand usage such as no common Iron or any other Iron but charcoal Iron perhaps could do, and it was for the latter that the Bowling Iron was originally manufactured as a substitute. Keeping in view the production of an uniform quality, changes of whatever description have been jealously regarded, and those that have been made have only been arrived at by very gradual stages.

Mr. H. W. Ripley, on behalf of the Bowling Iron Company, said that the works were started in the latter part of last century by three gentlemen, who each subscribed £5 for the purpose of testing the quality of the Iron ore which was found there. They had a few years ago expended over £12,000 in new engines, which had paid for themselves in eighteen months by the economy effected.

PARTICULARS OF THE LILLESHALL COMPANY'S WORKS, IN SHROPSHIRE.

Engineering Department.—In the machine shops there are upwards of 70 tools, consisting of lathes, screw-cutting and surfacing; large wheel lathes and face lathes; powerful planing machines; side planing machines for facing large cylinders and steam hammer standards; slotting, drilling, and shaping machines of various sizes; radial drills; cotter-hole drills; bolt and nut-screwing machines; cylinder-boring machine;

also vertical apparatus for boring cylinders up to 120 in diameter. Most of the tools are by eminent makers, and are of the newest designs.

The machinery is driven by a pair of wall engines and by two horizontal engines.

The erecting shops are spacious and well-lighted, and are so arranged with regard to the machine shops as to reduce to a minimum the labour of transporting the materials.

There are now in progress six locomotives, nine blast engines of various types, heavy pumping and winding machinery, steam hammers, &c.

In the smithy there are two steam hammers, one, recently-erected, being the American dead-blow ham mer. The smiths' fires are blown by a noiseless fan.

The foundry is specially arranged for the making of heavy castings, and is well supplied with cranes— swivel and overhead travelling. There are facilities for making castings up to thirty tons.

The fitting, erecting, and machine shops cover about an acre of ground.

General Remarks.—The Company raise annually about 400,000 tons of coal and slack, 105,000 tons of clayband ironstone, and 5,000 tons of fire-clay for bricks.

The number of persons employed in the mines, at the furnaces, and the engineering establishment is upwards of 3,200. This Company has nine blast fur naces, capable of making from 70,000 to 75,000 tons of pig iron per annum, of the very best quality.

[N]

IMPORTANT HINTS AND INFORMATION FOR IRONMASTERS
AND MERCHANTS.

The comprehensive List of Buyers of Iron, which commences at page 223, conveys valuable information.

It is just as well to inform the reader that the author is the Editor of the 'London Iron Trade Exchange,' the oldest metallurgical publication extant, the 'Mining Journal' excepted. The 'Exchange' was published as early as 1849, under the name of the 'Iron Trade Circular.' In March, however, of this year (1873), under the advice of Sir James Malins, one of Her Majesty's Vice Chancellors, the name was changed from the 'Iron Trade Circular' to the 'London Iron Trade Exchange,' edited by the Author. It has always been a subscription paper, and remains so, price 2*l.* 2*s.* per annum in advance; therefore, any Subscriber to the 'Exchange' is entitled to have all information from the Editor in respect to the Iron Trade, in answer to enquiries, free of all charge. Every name in the list to a Subscriber, might be made subject to our own reply in regard to respectability, standing in the City &c., &c.; besides, we may here remark that the Subscribers to the 'Exchange' have a right to ask for and receive information direct from the Editor upon all and every matter connected with the Iron Trade within the range of the Editor's own knowledge, free of charge;

THE DARLASTON STEEL AND RON COMPANY'S BLAST URNACES AND RON FOUNDR ES.

THE DARLASTON STEEL AND IRON COMPANY, LIMITED.

but it must be clearly understood that all answers in respect to standing and respectability of merchants must be treated confidentially, Mr. Griffiths' personal knowledge on the latter subject being varied and extensive, both in London and the Iron districts.

It may further be added that the letters on all subjects from Subscribers are answered promptly by Mr. Griffiths' personal dictation. The Editor takes pride in furnishing this information.

THE DARLASTON STEEL AND IRON COMPANY

Is one of the very oldest in South Staffordshire, and has always had a name for a good quality of iron. All sizes of iron are made here, including hoops, strip, tank plates, boiler plates, small sizes of rounds and squares, for the quality of which these works were noted in Bills' and Mill's times who established the works.

The Darlaston Steel Company likewise convert steel on a large scale on the cementation process. All kinds of steel are made here and the brands are well known and appreciated in the market. The company have three blast furnaces, extensive puddling furnaces, and rolling mills both here and at Wednesbury; also valuable iron-stone and coal mines, having large quantities of coal over and above their own consumption which are disposed of in the open market.

These extensive works were established by Bills and Mills. We believe they commenced about 1814, or earlier.   At the death of Mr. Samuel Mills, a company composed mainly of the Lloyd family purchased the concerns, adding to them the King's Hill Iron Works, well known for its boiler plates.

The Darlaston Steel and Iron Company have recently erected one of Brown's patent mills at considerable expense.   This is the only mill of the kind working in Staffordshire.   It is capable of rolling enormous quantities of strip of great lengths, the rolls being so arranged under Brown's patent as to complete the operation with very little manual labour, the piece being conducted by the power of the machinery into a second pair of rolls which completes the work.   They have likewise at work here two of Casson's patent puddling furnaces with Griffiths's puddling machine, which, we understand, work well together.   Casson's furnace is generally well thought of in this district.

This is one of the largest and most important concerns in Staffordshire, and as all the mines and coal in this locality are of superior quality, the Darlaston Steel and Iron Company have no difficulty in keeping up the high quality of iron always made at these works.

Mr. Sampson Lloyd, Wassel Grove, Stourbridge, late of the old Park works, is the Chairman of the company.   Mr. Francis Lloyd is the managing director.

This company has forty-three puddling furnaces, and seventeen reheating furnaces, three blast furnaces, three blast engines, seventy horse power each (altogether they have sixty-three steam-engines), eight rolling mills and a drawing-out forge for all kinds of use iron and

ONE OF THIS, AT DARLASTON.

ONE OF THE DARLASTON STEEL AND IRON COMPANY'S BLAST FURNACES, AT DARLASTON.

steel. They make rails, wire rods, and all other kinds of iron. Rails are laid down to all parts of the works. They have a self-tipping apparatus on the mine banks of the furnaces. Also extensive collieries and mines, 850 acres in all, 350 freehold and 500 leasehold, besides 55 acres of thick coal 12 yards thick. Their property contains the Brooch coal, the Heathen coal, and various other seams, the new mine stone, the rough hill white stone, the Gubbin and shales and blue flats.

This firm make channel iron and rails and sell in the open market very large quantities of coal. Besides their own pig iron which is of high Staffordshire quality, they purchase and consume as a mixture Lilleshall from Shropshire, and Barrow pig from Lancashire to mix with their own make. Name of managing director can be give at office, 84, Cannon Street.

The Albert and Moxley Iron Works.

The former of these works were built twenty-one years ago, and have since been carried on by Mr. David Rose. The Moxley Works, which were formerly carried on by Messrs. Daniel and David Rose, were founded in 1830, but on the occasion of Mr. Daniel Rose retiring from business a few years ago, the whole of the property was acquired by his youngest brother, Mr. David. These works are justly celebrated for the manufacture of use iron forgings and charcoal sheet iron.

The Victoria Works are famed for the manufacture of all kinds of strip iron, notably that for locomotive and boiler tube purposes.

The Albert Works, like Mr. William Rose of Batman's Hill, stand well in the market for plates and sheets, and the firm have a good old connection with engineers and machinists at home and on the Continent of Europe. Two blast furnaces have recently been built here on most modern principles, and acknowledged to be the finest plant in the South Staffordshire District, and capable of turning out 20,000 tons of pig iron per annum.

Mr. Rose has extensive galvanizing works here, and carries on the trade on the same premises to a large extent. Although Mr. Rose is an old ironmaster, his judgment in mines is very sound. On more than one occasion he has purchased valuable mineral property in the Black Country, and sold it at a large profit.

It may be interesting to note that near the works is a valuable sand mine, largely used for blast furnaces and mill furnace bottoms. It will thus be seen that whilst the sand in being excavated forms a nice little revenue, it is a valuable adjunct to the works for the deposit of cinders and ashes.

The mines of coal and ironstone here, and at an adjoining colliery of about 100 acres in extent, are also very prolific. Clay for the manufacture of fire-bricks, is also raised, and the quality is very superior. It is estimated that there is sufficient coal for the supply of the works for at least twenty years.

This is a most unique and valuable property, for Mr. David Rose, of Moxley, digs his own coal, sand and fire clay, makes his own pigs and fire-bricks, puddles his own iron, makes and galvanizes his own sheet iron, and, we believe, raises a large portion of the ironstone to make the pigs. We can safely say there are no other works in England, or the world, which can boast of the same products and advantages on one and the same spot, stretching over an area of comparatively only a few acres of ground.

The works, which are connected with the London and North-Western system, are entirely surrounded with a high brick wall, and are exceptionally couvenient and well laid out. In all there are 40 puddling and ball furnaces, 5 sheet mills, 1 plate mill, 1 strip mill, 1 bar mill, and one hoop mill. The pig iron made here is good. Indeed all the iron made at the Moxley Works, including their galvanized sheets, stands · high in the London market.

HISTORY OF PUDDLING, BEGINNING WITH HENRY CORT.

PUDDLING, in the sense used in connection with malleable iron making, requires the exercise of very trying manual labour. Both shingling and rolling test severely the muscular power and adroit manipulations of these iron-workers. Some think the collier and the miner have a trying and a severe physical task in the bowels of the earth. This may be so; we are of opinion, however, that the physical power and endurance exercised by the puddler to make a ' heat' of good iron is greater, and taxes the muscle and strength of the operator to a much greater extent than the shingler, the roller, collier, or any other workman engaged ، in the coal and iron trades, particularly if the puddler is ̨ bent upon doing his work well, so as to produce a proper yield and good iron. For the attainment of these *desiderata* good puddling is a *sine qua non*. We will not attempt to describe the process further than to say that for three-quarters of an hour the puddler has to face the molten metal, continually agitating the same in consecutive order over this boiling sea of metal and silica, which is so bright with the high state of calorific fluidity necessary for the successful process, and the workman being within a yard of the stopper-hole of the furnace, that and the meridian sun-like glare of the metal upon the eyes are almost overpowering. Nevertheless, he must work away until his iron comes into ' nature ; ' his exertions and the high temperature, particularly in the summer time, cause perspiration to such an extent that the puddler of necessity has a towel always ready to remove it. His final operation is to take the ' heat' in four, five, or six pieces, called ' balls,' previously formed in the 'furnace, to the hammer, where the iron is compressed and consolidated by heavy blows, which at the same time, drive the dross or cinder out of it, and in this way it is prepared for the rolls. The puddler now has breathing time; his work is light for an hour;

his only next duty is to assist in re-charging the furnace and quietly recuperate his strength for the next ' heat,' which will be melted ready for him in another hour or so.

The first puddling furnace was invented and constructed by Henry Cort about the year 1793, for which this pioneer iron-master took out a patent. His furnace was reverberatory ; the bottom was made of sand ; and although Cort's plan was very imperfect, and almost useless for making good iron, his invention of groove rolls about the same time, to co-operate with his new system of puddling, gave a great impetus to the iron trade in these kingdoms. The iron made from the pig, however, by Cort, was brittle, and wanted the fibre and strength which were afterwards supplied to the metal to some extent by the introduction of the refinery, which enabled the ironmasters of the United Kingdom, in a few years, to make sufficient iron for our own native consumption, which had previously been imported from Russia and Sweden at very high prices. The refinery, it may be remarked here, was a faint attempt in effect at Mr. Bessemer's wonderful process, although the inventor was ignorant of the causes which produced some good effect on the iron. The difference in the processes was great. The old refinery drove a blast of atmospheric air *on to* the molten iron for a short time after it was melted, which, to some extent, decarbonized it ; Mr. Bessemer, however, injected large volumes of atmospheric air at a much higher pressure into the very heart of the molten metal, and by this means ministered oxygen *ad libitum* to the utter destruction of all the carbon contained in the metal. Mr. Bessemer afterwards, by the addition of spieglesen. adds to the iron or steel, whichever is intended, the exact quantity of carbon required for the iron or steel, while in a molten state.

By the above it will be perceived that our fore-fathers, although ignorant of metallurgical art, and in the absence of scientific knowledge, accidentally developed a part of Mr. Bessemer's principle. This is a short description of the first puddling furnace and refinery, and although we can say but little in favour of this puddling furnace, Mr. Henry Cort left a legacy to the iron trade, in his invention of groove rolls, which cannot be too highly prized, and will always render the name of Henry Cort famous in the annals of the iron trade.

About the year 1825, sand-bottom puddling furnaces began to be abandoned in favour of what were then called 'iron bottoms.' Various silicates and oxides of iron were melted on the plates to make the bottoms and for fettling. This plan was adopted for some years, and to distinguish it from what we must call Mr. Joseph Hall's invention we shall call this the wet or water process. At a certain point of the heat, while the 'fore-hand' was puddling after the iron was melted, the 'underhand,' with a dish properly constructed for the purpose, threw water, *viâ* the 'stopper-hole,' upon the molten iron—about three half pints at once, the 'fore-hand' puddler frequently calling out, 'Water,

water,' as he thought it was required. What was called the drying process now commenced. The damper was dropped, and with the ' paddle' the puddler turned and returned the iron over in all parts of the furnace for some time, until the whole was properly disintegrated into a kind of powder, and was moved about until it became something like rough sand, though not so fine. Then the ' damper ' was pulled up, the word ' fire' was given by the puddler to the ' underhand,' the calorific power was increased rapidly, the iron became bright in the furnace, and, as its malleability and adhesiveness were gradually established, the puddler separated the iron, rolled it up into ' balls,' and with his tongs gaily dragged it to the helve. This was the old process.

The present boiling process is too well known to require a lengthened description here. It was discovered and invented by the late Mr. Joseph Hall, one of the partners at the great Bloomfield Works at Tipton, and from that day to this has been universally adopted, not only in the United Kingdom, but in all parts of Europe and America.

Briefly, Mr. Hall's process consists of *boiling*[1] the iron, not *roasting* it. Under the old plan the pigs lying on the bottom of the furnace were subjected to the action of the cutting flame passing over them more on the top part than the bottom of the charge of pig metal. On this account the upper portion of the pigs, particularly

---

[1] See Mr. Hall's book, which minutely describes this important invention.

the very highest part, became melted before the bottom ;
and as the action of the caloric was more intense upon
this portion of the iron by the passing flame, *ab initio*,
the puddler had . to watch the process, and, by means
of his paddle, remove the more extreme incandescent
portions, or turn the pigs over, in order that, by the
uniform application of the accumulating caloric, the
whole mass by these mechanical means might be rendered uniformly incandescent, and finally thoroughly
melted.   But during this irregular action of the fire on
the metal, one portion of the iron became liquid before
the other was thoroughly melted, and, as the heat
was constantly increasing, great waste occurred by
evaporation before the whole was properly melted, notwithstanding the greatest care on . the part of the
puddler.   Mr. Hall's process, however, introduced
boiling in silica.   Here the iron is protected by the
molten bath of silicâ which covers it, and although
more intense heat can be applied to the iron through
this molten sea of silica which envelopes it, the cutting
flame which continually passes over the furnaces spends
itself *upon* and is absorbed *in* the cinder and iron, without burning or damaging the iron.   While the iron is
in this boiling state it is the duty of the puddler to
agitate and puddle it, and, by all the means in his power,
to facilitate the elimination of the carbon, the exit of
which is seen plainly enough by the globules constantly
rising on the top of the molten cinder until the iron
comes into 'nature.'   This is, we believe, a correct,
though short description, of Mr. Joseph Hall's boiling

process ; for further particulars see Mr. Joseph Hall's book.

By this process uniformity in quality is obtained, waste is avoided to a considerable extent, and a quality of iron produced from the pig by the puddling process never attained by Ccrt's process or the intermediate 'iron bottoms' and water. It will be observed that the calorific power of the puddling furnace can only be kept up by the constant passage of volumes of flame over the iron in the well of the furnace. This necessitates a large consumption of fuel, and as coal is nearly treble the price it was, the cost of fuel now has become the great consideration, and unless some method can be adopted to economise the use of coal in puddling, it has become a grave question, in view of cheap coal in America, whether we shall be able to hold our position as the greatest iron producing country in the world. It will be observed that the flame—in fact, the whole calorific power—is constantly passing at a rapid pace over the iron, finally to be wasted and lost, for it merely leaves a portion of its calorific influence in the iron, and by the draught of the furnace is constantly carried into the open atmosphere and lost. Various patents have been taken out, the most important being Mr. Dank's, Mr. Siemen's, and Mr. Casson's. As the great object to be attained is to save labour and fuel, and make a superior quality of iron, whichever of these three patents is most adapted to produce these results must and will be most generally used, and confer the greatest benefit on the ironmaster, the puddler, and the world at large ; for it cannot be denied that the consumption of iron is so interwoven with the material and social progress of the age, that any amelioration in the cost of its production will be a blessing to mankind generally.

We went down to the Round Oak Works with a view of witnessing Mr. Casson's patent process, and have given the result of our observations for the information of our readers. The same account will likewise be found in the 'London Iron Trade Exchange,' No. 749, page 868.

THE BILSTON IRON COMPANY; THE FACTORY, DEEPFIELDS AND STONEFIELD IRON WORKS, BILSTON.

Messrs. Chambers and Sankey some years since commenced business, as most successful iron-masters have done before them, in a very small way at the Factory Works, but as years rolled on they have added the Stonefield Works and the Deepfields Works to the Factory : they therefore have the Deepfields Works, the Stonefield Works, and the Factory, which is thoroughly renovated and considerably enlarged. We ran through all their works, and were pleased with the arrangements of one and all.

At the Deepfields they have facilities for making sheets and boiler plates equal to any other works in the neighbourhood. The quality of the plates here is good, the sheets are well annealed, and the shears being of the most modern type, plates and sheets are turned out at the Deepfields, not only of very good quality, but in a clean and handsome condition. At the Factory works, sheets of thin gauges are made, including doubles and Satten, Canada plates, dish-plates, etc., the machinery being well adapted to this kind of iron.

At the Stonefield Works, hoops, bars, and small rounds and squares are manufactured.

This firm have in all thirty-two puddling furnaces and have a good name for the quality of the iron they produce, and are capable, we believe, of turning out as much sheet iron as any house in Staffordshire. They make Best Best Best and charcoal sheet iron, and all descriptions of boat and boiler-plates, Canada plates, dish-plates, chequered plates, hoop iron, bar iron, small rounds and squares, &c., &c. They raise their own coal. The partners are all practical Iron-masters, and give constant personal attendance and supervision over these important works.

We annex an engraving of one of the three works belonging to this firm hereto. It is the smallest of the three. All these works are situated in the very centre of the best coal district; each are within very few minutes' walk of a railway station. Head office at Stonefield Works.

THE BILSTON IRON COMPANY'S STONEFIELD IRON WORKS.

CHAPTER XXI.

WHICH has become quite indispensable for the Bessemer steel process, has been for a long time made at the iron-works of Schisshyttan, near Smedjebakken in Sweden, from magnetic iron ore containing 13 per cent. of manganese, with English coke, and, though fuel is very costly, the smelting pays well, the mine being close to the top of the furnace. This property last summer was transferred to the hands of some German capitalists. The principal locality, however, from which spiegeleisen is derived is the county of Siegen, in Prussia, where very fine steel ores, carbonates, and hydrates of Iron, with a large percentage of manganese, are produced at Stahlberg, Brüche, and Wilderman, near Müsen, Baud-enberg, Einigkeit, and Kunst, near Burbach, and Storch, and Schöneberg, Honigsmund, Eisenzeche, Alter Hamberg, Gilberg, Grimberg, Flossberg, Driesbach, Grauebach, and others near Siegen. All these ores are calcined in kilns before being smelted with charcoal, or coke, or both mixed, for spiegeleisen, steel pig, or Bessemer pig. When smelted for spiegeleisen the charge is composed of 1,700 lbs. of calcined Siegen ore, 600 lbs. red hæmatite, 1,000 lbs. limestone, and 1,500 lbs. coke or charcoal, and it is produced at an average cost of 5*l*. 15*s*. per ton of 1,000 kilogs. When steel pig for puddling steel is made from it, the burden consists of 1,150 lbs. of Siegen ore, 800 lbs. red hæmatite, 350 lbs. tapcinder, 1,000 limestone, and 1,250 lbs. coke, and it is produced at an average cost of 4*l*. 17*s*., and is a very clean white forge pig.   For the production of dark grey Bessemer pig the mixture is made of one-third Siegen ore, one third red hæmatite, and one-third specular iron ore, which is found in that country of excellent quality.   The charge is fluxed with a large quantity of limestone, say about 42 per cent., and requires from 100 lbs. to 200 lbs. of coke more per ton than spiegeleisen.   The smelting temperature is kept very high.   The pig Iron, when in a liquid state, shows on its surface. a peculiar change of groups of figures, which is the characteristic of spiegeleisen, and is the first indication of crystallisation. The slag is, outside, in a vitreous state of bluish or violet colour, with a more stony or crystalline interior of a yellowish brown tint.   The cost of producing Bessemer pig is 6*s*. to 7*s*. per ton less than for spiegeleisen.

Spiegeleisen is used in the manufacture of Bessemer steel, being added on the termination of the blowing process to supply the requisite quantity of carbon to the pure molten metal for its conversion into steel.   The Weardale Company and the Ebbw Vale Company manufacture this article.   The wants, however, of the converters in this country are mostly supplied by foreign makers.

The Weardale Company, ·we know, have most valuable deposits of Spathic ore, admirably adapted for its production, and produce a beautiful specimen of spiegeleisen, superior to any foreign yet introduced.   These are the only two firms in the United Kingdom able to make it.

CHAPTER XXII.

PIG IRON MANUFACTURED IN 1871.

IRON ORE.—The total quantity of Iron Ore raised in the United Kingdom amounted to 16,334,888 tons 14 cwt.

Foreign ores imported	.	.	.	.	324,034 tons.
' Burnt ore ' imported	.	.	.	.	200,000 ,,
			Total	.	524,034 ,,

Total of Iron ore returned as smelted in Great Britain 16,859,063 tons.

Value of the Iron ores of the United Kingdom, 7,670,572l.

Number of furnaces in blast, 673.

Pig Iron produced :—

In England	.	.	.	.	.	4,379,370 tons.
,, Wales	.				1,087,809 ,,
,, Scotland		.		.	1,160,000 ,,
Total production of pig Iron in Great Britain				6,627,179 ,,

This quantity, estimated at the mean average price at the place of production, would have a value of 16,667,947l.

Summary of Pig Iron produced in 1871.

Countries	No. of Iron Works Active	No. of Furnaces built in District	No. of Furnaces in Blast	Tons of Pig Iron made
ENGLAND.				
Northumberland	2	10	3	34,165
Durham	12	61	47½	759,244
Yorkshire, North Riding . .	17	75	70	1,029,885
„ West Riding . .	9	39	25	114,549
Derbyshire	12	46	38	270,485
Lancashire	7	41	34	520,359
Cumberland	8	34	28¾	336,569
Shropshire	19	25	19	129,467
North Staffordshire . .	7	35	30	268,300
South „ . .	53	163	18	725,716
Northamptonshire . . .	5	12	9	60,512
Lincolnshire	3	7	4	30,122
Gloucestershire . . .	} 6	18	13	99,997
Wiltshire				
Somersetshire . . .				
Total . .	160	566	429¼	4,379,370
WALES.				
NORTH WALES.				
Denbighshire	3	8	5	41,893
SOUTH WALES.				
Anthracite Furnaces . .	2	19	8	34,761
BITUMINOUS ⌜ Glamorganshire .	15	72	53	510,087
COAL ⌟ Brecknockshire .	1	14	4	30,086
DISTRICTS. ⌞ Monmouthshire .	11	62	47	470,982
Total . .	32	175	117	1,087,809
SCOTLAND.				
Ayrshire	7	43	35	
Lanarkshire	13	92	79	
Fifeshire	2	6	3	
Linlithgowshire . . .	2	6	4	
Stirlingshire	2	7	5	
Haddingtonshire . . .	1	1	1	
Argyleshire	0	1	0	
Total . .	27	156	127	1,160,000
Total as above . .				6,627,179

General Summary of the returns of the Mineral Produce of the United Kingdom received by the Mining Record Office for 1871.

Number of Mines	Mineral	Quantities		Value
		Tons	cwts.	£
2,760	Coal.	117,352,028	0	35,205,608
210	Iron ore [1] .	16,334,888	14	7,670,572
122	Copper ore	97,129	0	887,118
145	Tin ore	16,272	0	1,030,834
241	Lead ore .	93,965	17	1,155,770
47	Zinc ore .	17,736	10	56,330
33[2]	Iron pyrites (sulphur ores) .	61,973	0	64,987
1	Silver ore	5	0	421
16	Arsenic [3] .	4,147	15	15,519
9	Gossans, ochres, &c. .	697	5	1,396
1	Wolfram and tungstate of soda.	20	0	228
1	Nickel .	2	0	98
1	Bismuth .		2	14
2	Fluor spar	51	10	26
4	Manganese	5,548	1	22,958
1	Cobalt ore	3	0	120
	Barytes .	5,512	8	3,539
	Clays, *fine* and *fire* (*partly Estimated*).	1,255,000	0	475,000
	Earthy minerals (*Estimated*)			600,000
	Salt .	1,505,725	0	752,862
	Coprolites (*Estimated*) .	36,500	0	51,000
	Total Value of the Minerals produced in the United Kingdom in 1871. }			47,494,400

[1] It has not been possible, in every case, to determine whether the return of Iron Ore has been for *calcined* or *uncalcined* ores. The actual production of 'Raw Ore' will probably be in excess of this quantity. Estimating the quantity of pig Iron made at 2¾ tons of ore for each ton of Iron, and deducting the Foreign Ore, 'Burnt Ore,' and 'Cinder' used, the quantity will be about, or slightly above, 17,000,000 tons.

[2] Beside those mines, some collieries produced Pyrites, 'Coal Brasses.'

[3] Beyond this a quantity of Arsenic is produced by smelters, of which no return is obtainable.

IRON AND OTHER METALS SMELTED IN THE UNITED KINGDOM IN 1871.

Metals obtained from the Ores enumerated, &c., in the United Kingdom in 1871:—

Description of Metal		Quantities	Value
			£
Pig Iron . . . Tons		6,627,179	16,667,947
Copper . ,,		6,280	475,143
Tin . . . . ,,		10,900	1,498,750
Lead . . . . ,,		69,056	1,251,815
Silver . . . Ounces		761,490	190,372
Zinc . . . Tons		4,966	92,743
Other Metals (*Estimated*) .		—	3,000
Total Value of Metals produced from the Ores of the United Kingdom in 1871 .			20,179,770

Total Value of the Metals produced which are not smelted, Coal and other Minerals raised in 1871:—

	£
Metals, value of as above	20,179,770
Coal ,, ,, ...	35,205,608
Minerals, earthy, &c.	1,936,515
Total value	57,321,893

In this gross sum of 57 millions sterling, neither building stones, lime, slates, or common clay and brick earths are included.

IRON ORE PRODUCE.

General Summary of Returns.

Counties, &c.	Quantities	Value
	Tons cwts.	£ s. d.
Cornwall . . .	21,947 14	11,509 8 3
Devonshire . . .	14,124 14	6,095 16 0
Somersetshire . . .	32,883 13	32,883 13 0
Gloucestershire . .	207,598 16	155,060 0 0
Wiltshire . . .	159,894 0	27,989 0 0
Oxfordshire . .	28,330 0	21,247 0 0
Northamptonshire .	779,314 3	133,155 18 0
Lincolnshire . . .	290,673 9	50,705 0 0
Shropshire . .	415,972 0	157,083 0 0
Warwickshire . .	34,075 0	15,570 0 0
Staffordshire, North .	1,513,080 0	921,022 0 0
,, South .	705,665 0	453,512 0 0
Derbyshire . . .	492,973 0	295,782 0 0
Lancashire . . .	931,048 0	1,163,810 0 0
Cumberland . . .	1,302,703 15	1,448,975 0 0
Yorkshire, North Riding .	4,581,901 0	1,144,974 2 0
,, West Riding .	407,997 0	101,998 10 0
Northumberland and } Durham . . }	285,297 0	71,349 0 0
NORTH WALES . .	51,887 0	23,348 0 0
SOUTH WALES and } MONMOUTHSHIRE }	969,714 10	543,422 16 0
ISLE OF MAN . .	75 0	37 10 0
SCOTLAND . . .	3,000,000 0	825,000 0 0
IRELAND	107,734 0	66,043 0 0
Total Iron Ore Produc- } tion of the United } Kingdom . . }	16,334,888 14	7,670,572 13 8
'Burnt Ore' from Cup- } reous Pyrites }	200,000 0	——
Iron Ore Imported .	324,175 0	——
Total of Iron Ore of } which returns have } been received . }	16,859,063 14	——

CHAPTER XXIII.

OUR COAL IN THE UNITED KINGDOM.

THE Gross Quantity of Coal raised in the United Kingdom in 1872 *was* 121,000,000 *Tons.*

The coal mines of England are, without doubt, the most valuable indigenous product we can boast of, leaving any other mineral or metal produced in this country far behind in the quantity raised or their aggregate value.

Germany, Belgium, and France have coal, with extensive collieries regularly at work. The quality in these countries, in the main, is inferior to English, particularly for Iron smelting.[1] The export to Germany and France from this country has steadily increased over the last ten years, and during 1872 in an accelerated *ratio.*

The same may be said, in a lesser degree, with respect to all foreign countries depending upon England for their supply of fuel. We raised in 1872, 121,000,000 tons of coal and slack in the United Kingdom. According to Mr. Robert Hunt's valuable statistics the out-put of 1871 was 117,352,028 tons.

Mr. Hunt puts the gross value of 1871 at 35,201,618*l.*;

---

[1] The coal deposits in the United States may be said to be inexhaustible ; the same may be said, though in a lesser degree, with regard to the Dominion.

taking the 121,000,000 tons raised last year at 11s. 3d. per ton in trucks, boats, or F.O.B., which we think will be quite low enough, our coal for the year yielded to the coal-masters no less than 68,625,000l., which far supersedes in value the tea of China, the cotton of America, the wines of France, the diamonds of Golconda, or the native produce of any single article in any other country in more southern climes, where the genial influence of the sun so materially assists the labour of man in bringing forth corn, wine, and the more luscious fruits of the earth. Cornwall has ministered from the earliest ages to England's requirements for tin and copper. Shropshire, Northumberland and Durham, Montgomeryshire and the Isle of Man, have distanced Cornwall in the out-put of lead, and the totals of these metals produced in the United Kingdom are highly creditable to the adventurers, but the aggregate value of all the metals raised appear insignificant when placed in contrast with the gross value of our '*Black Diamonds.*'

The value of the tin smelted from English ore in 1871 was 1,498,750l.; the value of the lead was 1,251,815l.; the copper was 475,143l.; the silver 190,372l.; zinc 92,743l., which gives a total of 3,508,823l. If for the advance in the price of copper and tin, which ascended most in value in this category last year, we add 500,000l., which will be over the mark, this gives us a total value for all the metals for 1872 of 4,000,000l., not a 1-16th, or one and threepence in the pound, of the total value of the coal raised in the United Kingdom; the out-put of which, as before stated, has produced to the coal-owners

no less than 66,000,000*l.* sterling. Our gracious
Queen may justly boast of the gold of Australia,
the precious diamonds of the Cape, the indigo and
opium, and of late the cotton and jute, of India;
but her vast dominions, either at home or abroad, are
unable to furnish any single staple commodity or manufacture, which can compete in value and national importance with the coal annually raised in the United
Kingdom. The rapid and marvellous 'jumps,' as Mr.
Gladstone so cogently puts it, in the acquisition of
national wealth, must be attributed in no small degree
to our ability to furnish a constant supply to the world's
increasing demand for this valuable mineral. It
generates our steam, applied first successfully by
Watt's condensing engine, to give motion to the
wheels and spindles of corn-mills and factories, and
ten thousand other manufacturing processes pro-
pelled by the steam-engine. Steam, through the engine, is manipulating, day and night, in the Potteries,
reducing bones and the hardest flint into snow-white
china clay; in Manchester and Leeds, driving hundreds of thousands of spindles at speeds which would
bewilder the reader to compute; indeed, steam is
everywhere employed as it were to civilize and edu-
cate inanimate nature itself, converting the sand of the
sea-shore into a beautiful button, the hard feltspar
rock into a china teacup, and the cotton of Georgia into
dresses and comfortable under-linen for the use of
mankind. Coal likewise supplies the motive power to
the railway, enabling us to move from one end of the
kingdom to the other with extraordinary rapidity.
Again, our gas owes its lighting power to coal, which

enables us to light up our towns, and contravene the inexorable laws of Nature by turning night into day, while the earth, during her diurnal rambles, for a little while would fain leave us without the smiles of the sun to light, warm and cheer us. Coal propels, through the engine, the mighty ironclads which protect our island ; these are ready, by the power of steam, to dash into any port or harbour in the world to avenge our outraged honour or demand redress for England's wrongs at the hands of her ene-mies. Coal is the handmaid of trade and commerce ; by the steam it creates our ships easily plough their way through the mighty deep to all parts of the world, gallantly steaming back into our great ports, laden with the richest and most choice products of all nations, making this country at once the *depôt* or grand empo-rium of the gold and all other produce most valued and eagerly sought after by all civilized nations.

After endeavouring to represent the vital importance of English coal to our readers, with all its surroundings in a national and commercial point of view, we will now explain and name all the counties in England, Scotland, and Wales, whence the aggregate out-put was raised last year, giving the amount for each county, and showing the degree in which they contributed to the total aggregate.

First, however, we give Mr. Robert Hunt's, F.R.S., general summary for 1871, the returns having been obtained by that gentleman in his official capacity of Keeper of Mining Records, the statistics having been published by order of the Lords Commissioners of Her

Majesty's Treasury ; therefore, these 1871 returns are as perfect as it is possible to get them for the present. A careful perusal of these tables, and comparison of the out-put for 1871, both in coal, Iron, and Iron ore, the reader will be enabled to judge for himself of the estimate which we have given at the end of the tables of the out-put for 1872, which has been considerably increased.

LANCASHIRE.

NORTH AND EAST OR MANCHESTER DISTRICT.

Mr. Joseph Dickinson, Inspector.

The Number of Collieries, 287.

The production of coal . . . . . 7,576,000 tons.

THE WESTERN DISTRICT.

Mr. Peter Higson, Inspector.

The Number of Collieries, 89.

The production of coal . . . . . 6,275,000 ,,

Total of Lancashire . . 13,851,000 ,,

CHESHIRE.

Mr. Thomas Wynne, Inspector.

The Number of Collieries, 29.

The production of coal . . . . 975,000 tons.

YORKSHIRE.

Mr. Frank N. Wardell, Inspector.

The Number of Collieries, 423.

No. of Collieries.		No. of Collieries.	
4	Bingley.	31	Dewsbury.
44	Barnsley.	29	Halifax.
49	Bradford.	26	Huddersfield.

No. of Collieries.		No. of Collieries.	
12	Holmfirth.	25	Rotheram.
99	Leeds.	36	Sheffield.
6	Normanton.	48	Wakefield.
6	Peniston.	1	Saddleworth and Settle.
7	Pontefract.		

Total of collieries, 423.

Total produce of Yorkshire, 12,801,260 tons.

NORTHUMBERLAND AND DURHAM.

Collieries.

Mr. George W. Southern, inspector, Northumberland 164
Mr. James Willis, inspector, Durham . . 140

Total collieries . 304

*Gross Produce sold from and used at the Collieries of Durham
and Northumberland in the year* 1871 ·—

Tons.

Coal exported to foreign countries 6,230,567
Coke exported to foreign countries 289,314
tons, computed as coal 482,190
Coal sent coastwise 5,355,737
Coke sent coastwise, 19,005 tons, computed
as coal 31,675
Coal carried from this coal-field by railway 6,227,002
Coke carried from this coal-field by railway,
for local and land sale as below, 2,161,020
tons, computed as coal 3,601,700
Coke carried south of Altofts, 1,182,160
tons, computed as coal 1,970,266
Coal and *coke* for railway use, the coke
computed as coal . . . 591,779
Colliery consumption *estimated* . 1,450,000
Domestic consumption, and coal used in
local manufactures 3,250,000

Total of Durham and Northumberland . 29,190,916

CUMBERLAND.

Mr. George W. Southern, Inspector.

The Number of Collieries, 27.

The production of coal, 1,423,661 tons.

DERBYSHIRE, NOTTINGHAMSHIRE, LEICESTER-SHIRE, AND WARWICKSHIRE.

Mr. Thomas Evans, Inspector.

The Number of Collieries, 187.

The production of coal, 9,252,900 tons.

No. of Collieries.		Quantity produced. Tons.
130	Derbyshire	5,360,000
27	Nottinghamshire	2,469,400
18	Warwickshire	723,600
12	Leicestershire	699,900
187	Total of above counties	9,252,900

MONMOUTHSHIRE, BRECKNOCKSHIRE, and the edge of GLAMORGANSHIRE.

Mr. Lionel Brough, Inspector.

The Number of Collieries, 74.

The production of coal, 4,915,525 tons.

GLOUCESTERSHIRE AND SOMERSETSHIRE.

Mr. Lionel Brough, Inspector.

The Number of Collieries, 101.

The production of coal, 2,086,475 tons.

FOREST OF DEAN.

Coal worked in the year 1871, 837,893 tons.

Total produce of Gloucestershire and Somersetshire :

No. of Collieries.		Tons.
66	Gloucestershire	1,412,597
35	Somersetshire	673,878
101		2,086,475

WALES.

NORTH WALES.

Mr. Peter Higson, Inspector.

The Number of Collieries, 58.

The production of coal :

No. of Collieries.		Tons.
32	Flintshire	
24	Denbigshire	2,500,000
2	Anglesea	
58		

SOUTH WALES.

Mr. Thomas E. Wales, Inspector.

The Number of Collieries, 299.

No. of Collieries.		Tons.
9	Pembrokshire . .	111,000
44	Caermarthenshire	679,322
246	Glamorganshire	8,329,678
299	The production of coal	. 9,120,000

STAFFORDSHIRE AND WORCESTERSHIRE.

NORTH STAFFORDSHIRE.

Mr. Thomas Wynne, Inspector.

The Number of Collieries, 115.

The production of coal, 4,300,000 tons.

CONSUMPTION AND DISTRIBUTION.	Tons.
Coal used at Iron works	1,825,000
Coal used at potteries and brick works	760,000
Coal used at other manufactories .	525,951
Coal carried by N. Stafford railway out of the district	241,841
Coal carried by canal (Trent and Mersey) navigation	356,964
Coal carried by railway to local stations for home use	505,244
Colliery consumption	30,000
Total for North Shields .	4,250,000

SOUTH STAFFORDSHIRE AND WORCESTERSHIRE.

Mr. James P. Baker, Inspector.

The Number of Collieries, 307.

The production of coal, 10,031,250 tons.

CONSUMPTION AND DISTRIBUTION.	Tons.
Coal used in Iron works . .	3,585,750
Coal used by other manufactures	1,500,000
Domestic consumption	1,875,500
Colliery consumption and allowance coal	1,350,000
Total used in District	8,311,250
Sent out of district by railway and canal .	1,720,000
Total produce of South Staffordshire	10,031,250

Total coal produce of Staffordshire and Worcestershire :—

	Tons.
North Staffordshire	4,250,000
South Staffordshire and Worcestershire .	10,031,250
Total .	14,281,250

SHROPSHIRE.

Mr. Thomas Wynne, Inspector.

The Number of Collieries, 59.

The production of coal, 1,350,000 tons.

SCOTLAND.

THE WESTERN DISTRICT.

Mr. William Alexander, Inspector.

The Number of Collieries, 204.

	Tons.
The production of coal . . .	6,554,365

THE EASTERN DISTRICT.

Mr. Ralph Moore, Inspector.

The Number of Collieries, 216.

The production of coal	8,883,926
Total	15,438,291

Names of Inspectors	District under Inspection
George W. Southern, Esq., 17, Wentworth Place, Newcastle-on-Tyne.	Northumberland, Cumberland, and Durham North of the Wear.
James Willis, Esq., Old Elvet, Durham.	Durham South of the River Wear in its course from the sea at Sunderland up as far as Harraton near Chester-le-Street, and from thence westward, the line of the Ponton and Shields branch of the North Eastern Railway.
Frank N. Wardell, Esq., Wath-on-Dearne, near Rotherham.	The West Riding of Yorkshire.
Thomas Evans, Esq. Field Head, Belper.	Derby, Nottingham, Warwickshire, and Leicester.

Names of Inspectors	District under Inspection
Thomas Wynne, Esq., Stone.	North Staffordshire, Shropshire, and Cheshire.
James P. Baker, Esq., Tattenhall, Wolverhampton.	South Staffordshire and Worcestershire.
Joseph Dickinson, Esq., Pendleton, Manchester.	North and East Lancashire, called the Manchester District.
Peter Higson, Esq., Brooklands, Swinton, Manchester.	West Lancashire, the Wigan and St. Helen's Districts, and North Wales.
Lionel Brough, Esq., 11 West Mall, Clifton, Bristol.	Monmouthshire, Gloucestershire, Somersetshire, and Devonshire.
Thomas E. Wales, Esq., Brunswick place, Swansea.	South Wales coal field.
Ralph Moore, Esq., 7, Queen's Square, Glasgow.	SCOTLAND.— *Eastern Division*—including East Lanarkshire, Fifeshire, Clackmannanshire, Haddingtonshire, Edinburghshire, Linlithgowshire, East Stirlingshire, &c.
William Alexander, Esq., 23, India Street, Glasgow.	SCOTLAND. — *Western Division*—including Ayr, Dumfries, Dumbarton, West Division of Stirling, and part of Lanarkshire.

MINERAL STATISTICS FOR THE YEAR 1871.

SUMMARY OF COAL PRODUCE OF THE UNITED KINGDOM FOR 1871.

	Tons.
Durham and Northumberland . . .	29,190,916
Cumberland	1,423,661
Yorkshire	12,801,260
Derbyshire	5,360,000
Nottinghamshire	2,469,400
Warwickshire	723,600
Leicestershire	699,900
Staffordshire and Worcestershire	14,281,250
Lancashire	13,851,000
Cheshire	975,000
Shropshire	1,350,000
Gloucestershire.	1,412,597
Somersetshire	673,878
Monmouthshire	4,915,525
South Wales	9,120,000
North Wales	2,500,000
Scotland	15,438,291
Ireland	165,750
Total produce of the United Kingdom .	117,352,028

We estimate the coal raised in 1872 in the United Kingdom at 121,000,000 of tons, the Iron ore raised at 19,000,000 of tons, and the quantity of pig Iron made in the United Kingdom 7,250,000 tons.

After well considering the matter, we see good reason for giving the above figures, which will turn out to be as near the mark as can be expected in a mere estimate.

o

We have received returns, but as these are not complete, we have been obliged to fill up the hiatus according to our own judgment, which, relying on the data in our possession, has been exercised with care, after much consideration.

Scotland shows a falling off. Cleveland and most other districts exhibit an increase in the make of Iron in 1872. Coal and Ironstone have been consumed in the same increased *ratio*. The great activity over last year in all trades and manufactures consumed more coal by far than formerly.

The export demand was insatiable all the year; we therefore think that 121,000,000 tons of coal is as low as we ought to reckon the aggregate out-put for the United Kingdom.

The area of the British coal-fields is estimated at 4,250,000 acres, and no less than 64,661,000 tons of coal were raised in the United Kingdom in the year 1854, 20 years since valued then àt 14,975,000*l*. Of this sum 47,422,000 tons were raised in England, 9,643,000 in Wales, and 7,448,000 in Scotland, and 148,000 in Ireland. The area of the coal-fields of France is about six times less than that of Great Britain, whilst their product is sixteen times less. The total number of collieries in the United Kingdom in 1854 was 2,327. Of these 1,704 were situated in England, 306 in Wales, and 368 in Scotland, and 19 in Ireland. The average number of tons raised to each workman employed in the United Kingdom in 1854 was 293; in France in 1852 the average was only 136 tons to each workman engaged

NUMBER OF COLLIERIES IN EACH INSPECTOR'S DISTRICT.

ENGLAND AND WALES.

District	Name of Inspector	Number
Northumberland and Durham, north division . Cumberland	Geo. W. Southern, Newcastle-on-Tyne.	164
Durham, south division	James Willis, Old Elvet, Durham.	140
Yorkshire . . .	Frank N. Wardell, Rotherham.	423
Derbyshire . . Nottinghamshire Warwickshire Leicestershire	Thomas Evans, Belper.	187
Cheshire . Shropshire . . Staffordshire, North	Thomas Wynne, Stone.	203
Staffordshire, South, and Worcestershire .	James P. Baker, Wolverhampton.	307
Lancashire, North and East, or the Manchester district .	Joseph Dickinson, Pendleton, Manchester.	287
Lancashire, St. Helen's and Wigan Flintshire . Denbighshire Anglesea .	Peter Higson, Manchester.	157
Gloucestershire . Somersetshire and Devonshire . Monmouthshire . East of Glamorganshire . .	Lionel Brough, Clifton, Bristol.	196
Glamorganshire . Pembrokeshire . Caermarthenshire	Thomas E. Wales, Cae Bailey, Swansea.	246

NUMBER OF COLLIERIES IN EACH INSPECTOR'S DISTRICT—*continued.*

SCOTLAND.

District	Name of Inspector	Number
Lanarkshire, west division . . Ayrshire . . . Stirlingshire, west division . . Dumbartonshire Renfrewshire Argyleshire . Dumfriesshire	William Alexander Glasgow.	204
Lanarkshire, east division . . Fifeshire . . Clackmannanshire Haddingtonshire . Kinross-shire . Edinburghshire . Linlithgowshire . Stirlingshire, east division . . Peeblesshire. Perthshire . .	Ralph Moore Glasgow.	216

IRELAND.

District	Name of Inspector	Number
Ulster coal field Connaught . Leinster coal field Munster coal field	No inspector.	30
	Total for United Kingdom .	2,760

COAL AND COKE.

All the Countries receiving the Principal Exports in the Year 1871.

Countries to which Exported.	Quantities Exported.		Declared Value.	
	Coal.	Coke.	Coal.	Coke.
	Tons.	Tons.	£	£
Russia : Northern Ports . .	665,360	40,099	314,168	26,412
„ Southern Ports . .	207,228	1,101	102,141	813
Sweden	369,666	27,884	173,090	20,145
Norway	221,655	7,102	95,230	5,000
Denmark	648,191	10,510	284,119	7,495
Germany	2,331,304	65,507	969,839	42,965
Heligoland	116	—	40	—
Holland	502,055	4,310	223,292	3,087
Belgium	116,703	50	50,443	30
Channel Islands .	74,302	568	36,294	394
France . .	1,968,227	8,971	889,049	4,898
Portugal, Azores, and Madeira	169,631	4,228	87,748	3,215
Spain and the Canaries . .	475,852	78,966	257,698	56,281
Gibraltar	131,331	33	74,584	26
Italy	791,897	16,498	366,906	10,962
Austrian Territories . .	85,016	7,104	41.740	4,865
Malta	186,957	10	102,976	11
Greece (including Ionian Islands)	61,704	17,351	33,942	11,294
Turkey	277,004	1,505	146,833	1,103
Wallachia and Moldavia .	38,704	224	18,540	135
Egypt	451,912	3,557	230,317	2,318
Tripoli and Tunis . . .	1,995	—	1,038	—
Algeria	25,219	58	12,368	34
Morocco	629	—	326	—
Western Coast of Africa .	49,289	.	29,298	—
Ascension	3,065	.	1,712	.
St. Helena	150	—	150	—
British Possessions in South Africa	20,792	1,642	12,130	1,357
Eastern Coast of Africa .	1,768	—	1,020	—
Mauritius	13,748	62	7,346	98
Arabia, Aden . .	87,394	100	47,866	100
„ Muscat .	1,426	—	855	—
Persia	602	.	602	—
British India : Continental Territories . .	335,210	15,915	187,392	16,405
„ Straits Settlements .	114.373	—	66,206	—
„ Ceylon . . .	107,625	138	57,479	170
Java	44,840	924	24.406	1,146
—— Other Dutch Possessions.	3,432	—	1,736	.
Phillipine Islands . . .	2,580	—	1,621	—
Borneo	501	—	251	—
Siam	1,302	125	1,331	167

All the Countries receiving the Principal Exports in the Year 1871 *(continued).*

Countries to which Exported.	Quantities Exported.		Declared Value.	
	Coal.	Coke.	Coal.	Coke.
	Tons.	Tons.	£	£
Cochin China	650	—	392	—
China and Hong Kong . .	90,575	137	55,192	187
Japan	14,083	30	8,383	36
Australia . .	8,711	924	6,128	1,214
Islands in the Pacific .	480	—	251	—
British North America .	189,274	1,406	86,318	996
United States of America :				
On the Atlantic . . .	91,483	.	61,524	-
On the Pacific . . .	60,365	.	31,596	-
British West Indies .	175,335	224	99,387	308
Foreign West Indies .	281,877	452	149,574	335
Mexico	2,821	.	1,227	. .
Central America . . .	114	30	114	47
United States of Columbia (New Granada) . . .	11,241	—	7,190	. .
Venezuela	370	5	204	5
Ecuador	1,015	—	896	—
Peru	109,393	2,137	70,410	1,605
Bolivia	2,094	.	920	.
Chili	101,203	4,146	48,734	2,928
Brazil	316,417	12,890	188,036	9,083
Uruguay . . .	96,648	303	65,888	228
States of the Argentine Confederation . . .	62,860	4,639	42,970	3,521
Falkland Islands . . .	245	—	224	—
Total . . .	12,208,009	341,865	5,879,680	241,419

CHAPTER XXIV.

GLASGOW affords great facilities for shipping the Iron made in that district, the Clyde coming up to the City and ministering to Connal's great stores in the most convenient manner. The shipbuilding yards on the banks of the Clyde are the largest and most numerous in Great Britain. Steamers of all sizes are built here for all nations; the skill of the Scotch engineers is unrivalled in England, and the names of Napier, Thomson's, and Elder & Co., are well known in connection with high class steamers in all parts of the world. The best steamers in the world are built on the Clyde. This noble river, with hundreds of building yards on both banks for miles, as you steam out of Glasgow, presents one continued succession of gigantic works, occupied principally in Iron naval architecture. Steamers are launched every month, and the great works themselves are mementoes not only of the superiority of the Clyde makers over all others, but as a proof of the grandeur of Scotch enterprise, of which England has good reason to be proud.

Mr. John Elder has done wonders in carrying out the wholesome advice so often and ably given by Sir William Armstrong, to economize fuel in the generation of steam by perfecting the high and low pressure steam-engine, which is *now* accepted by the great engineers on the Clyde. These engines, at this time, are universally applied at with from sixty to eighty pounds pressure in the direct acting compound surface condensing engine, with a saving of just one half the quantity of fuel; the present firm of John Elder & Co. are still pursuing the same course of economy, with further improvements. This firm, by adopting Rowan & Orton's (of Glasgow) Patent Boiler, declare that with a pressure of 120 lbs. and three cylinders, they will reduce the consumption of fuel to a minimum of 1 lb. of coal per horse-power per hour. Steamers are built here with 3,000 horse-power each, and Messrs. Caird & Co. have engines in hand now for one vessel 5,220 horse-power. Mr. Robert Duncan, Port Glasgow, is now engaged on plans of steamers 600 feet long, probably to trade between the great Ramsden dock at Barrow and the port of New York. This noble river, with its engine-shops and yards, is a marvel in the metallurgical industries of our land; and by improved steam-engines and noble Iron steamers will, with honour, transmit to posterity the names of John Elder, Laird & Co., Robert Duncan, the Elders, Napiers, Dennys, and Sir William Thompson. Rankine is gone, and the Clyde loses his transcendant ability; but John Elder has made munificent provision for instructing suitable young men who aspire to practise the arts of building Iron ships and steam engines, and practically to apply those great principles which Rankine taught so well.

CHAPTER XXV.

THE NEW TRADE OF STEEL CASTING

HAS made rapid progress in this country since its first introduction at Krupp's great works; and as cranks, and all other important parts of steam engines and other machinery, can now be run out from the crucible of the best material, where tensile and tortuous strains are excessive this kind of manufacture is invariably preferred.

CRUCIBLE STEEL CASTINGS

Is a branch of the Sheffield steel trade of modern development, and promises to become of great magnitude. Krupp, of Prussia, was one of the first to make large castings in steel, but Krupp has long since been surpassed by Messrs. Vickers, of Sheffield, who may be considered the founders, &c., by originating and manufacturing all kinds of steel bells, and who are still the only large makers of crucible castings of steel, in large contracts, in England. To show the increasing importance of this trade, Messrs. Hadfield's Steel Foundry Company, Newhall, Sheffield, have just completed extensive premises and plant, covering a couple of acres with machinery, and improved appliances, for the manufacture of smaller steel castings of all descriptions, and who claim to be unsurpassed. The firm devote their personal attention, and claim a special excellence in having secured sound castings in crucible steel, from 1 lb. to several tons in weight, and also a superior method of annealing their castings to any engine, their success being such as to necessitate the erection of their new works to meet their increasing trade. We are informed by practical parties that they possess one of the most complete and well-arranged foundries of the kind. It is really surprising to learn how many and varied are the uses these castings are applied to, from reaping machine fingers to screw propellers for steam ships. This seems a great step, but from hydraulic cylinders to colliery, tram, and railway wheels seems greater. Pinions, engine shafts, plough-shares, horn-blocks, and axle-boxes, will give some idea of the articles to which they are being applied. In fact, crucible steel castings are fast replacing metal work and wrought Iron forgings, their superior qualities, viz., great tenacity, strength and lightness, giving them special advantages over other metals.

We understand Mr. R. Hadfield has patented an improved double disc railway wheel, the tyre of which is steel or Iron, with a metal centre *welded* thereto ; owing to its construction, it is a certain preventative against all railway accidents from tyres breaking. It can be manufactured cheaper than ordinary wheels. Mr. H. has also patented a capital cheap and durable method of fastening crucible steel wheels fast on the axles, so that they cannot work loose. We understand that

Messrs. Hadfield's Steel Foundry Company recently received from one firm alone an order for fourteen hydraulic steel cylinders, weighing upwards of one ton each. This will give some idea of their capabilities.

CHAPTER XXVI.

WILLENHALL AND ITS LOCKS AND BOLTS.

WILLENHALL, the real seat of the lock, door-bolt, and latch manufacturers for the world, is a township in the parish of Wolverhampton, and is connected with it by two lines of railways, viz. the old Grand Junction line, and one recently opened, the Wolverhampton and Wallsall line. Willenhall being just three miles from either town, now contains about 20,000 inhabitants, a complete hive of industry. We believe there are some five to six hundred separate manufactories (of course some only small concerns) of rim, mortice, drawback, dead, cupboard, drawer, box, and pad locks; all kinds of latches and door bolts; currycombs, grid-irons, box-iron stands, and skewers; horse scrapers and singers; carpet-bag frames and locks; box corners and clips; keys of all descriptions; and stampers of an endless variety of articles for the gun, steel, toy, and other trades carried on in neighbouring towns. There are also Ironfounders, brassfounders, and wrought Iron works, blast furnaces and collieries in abundance. To these mainly must be attributed the rapid growth of this industrious town. The writer can well remember when three to four thousand was the extent of its population. In those happy old days of the past there was one church, with a blaspheming drunken parson, who spent six times more of his time in the public-house than in the church, the only one the place possessed. In those times there was no Methodist or Dissenting resident minister; and what is more, no magistrate, no lawyer, no police, and not an inhabitant (except the parson) but what was engaged in some kind of business. At the present time the township of Willenhall contains four churches, five Wesleyan chapels, four Baptist chapels, five Methodist chapels of various denominations, and one Roman Catholic chapel, which represents one place of worship for every thousand of the population, a fact few towns can boast of; with good school accommodation, British, National, and Wesleyan, and a literary institute of no mean pretensions, having its reading, recreation, and class rooms, a good lecture hall, and a well-furnished library.

On visiting some of the manufactories of Willenhall we found the Albion works one of the most prominent, employing some hundreds of work-people. The business carried on here was established in the last century by the father of one of the present proprietors; and one of the principal branches of the trade, that of door bolts, was extensively carried on by the grand-father of the other more than eighty years ago. At these works we find manufactured rim, dead, and mortice locks; spring, rim, night, Norfolk, Suffolk, and Lancashire thumb-latches, in various ornamental designs; door-bolts in prodigious quantities. When in full employ they can produce 120 gross per week of this one article, aided by steam stamps, Nasmith's steam hammer, and steam presses for forming the various parts of the bolt, and piercing the holes in the plates; also pulleys for every conceivable purpose; sash, signal, and sliding doors; hat and coat hooks; door buttons, and black ironfoundery generally. We also find a large quantity of door-lock knobs, sold principally for export, called 'Harper & Co.'s patent mineral lock furniture.' At the time we write, they are engaged on orders for nearly 100,000 brackets, made of malleable cast Iron—these are for various telegraph lines, both for home and abroad. In this branch the Messrs. Harper & Co. excel. Our pen would fail to write the variety of purposes to which the malleable castings are now applied; suffice it to say, their patterns consist of more than *three thousand* different kinds and sizes

The largest lock manufactory here, and perhaps one of the oldest, is Carpenter & Tildesley's, well-known in England and all the Colonies for their locks of various kinds, particularly 'rim' locks. They are the largest lock and curry-comb makers in Willenhall, and employ the greatest number of hands in lock and key making of any house in Willenhall, having especial machinery for this purpose. This firm stamp their own keys. Mr. James Tildesley is son-in-law to the late Mr. Carpenter, and is the proprietor of these ingenious works. This is a highly respectable firm, and capable of executing orders to any extent of all kinds of locks.

CHAPTER XXVII.

WE know of no concern which has done more to
raise the character for engineering skill in the Black
Country than Thomas Perry & Sons. This is a highly
respectable old firm engaged in metallurgical pursuits.
50 years since—the firm then was Thomas Perry—
the family resided at Bilston, and as there were at
that time numerous Perrys in business, Mr. Perry, to
distinguish him, was always called 'Gentleman Perry,'
and was highly esteemed and loved by all classes.
Mr. Perry, with Mr. Sparrow, was one of the founders
of the Bilston District Bank, and remained a director
until his death.

He was quiet, unostentatious, always a gentleman,
was never known to forget himself or become agitated
in business; he was amiable and polite to all; a great
friend to his church and the clergy; a man of spotless
and the highest reputation. In the social circle at
home for goodness he was unequalled in all Stafford
shire; so amiable, so gentle, truly polite and kind
was Mr. Perry in his own family to each member
to the day of his death, although he lived to an advanced age. Mr. Charles T., the eldest son, married

Harriet, the third daughter of the late G. B. Thorney-croft, Esq., the first mayor of Wolverhampton and the founder of the firm of G. B. Thorneycroft & Co. Mr. Thomas Perry, the second son, is now at the head of the concern under review.

This foundry has a world-wide fame, and has done more in supplying new machinery and appliances, abreast with the rapid progress the Iron trade has made during the last ten years, than many other establishments.

Steam engines and all kinds of machinery are their great specialities, with all modern inventions in rolls and machinery in Ironworks. At this foundry the choicest brands of pig Iron are kept in large quantities, including Lilleshall, Madeley Wood, Blaenavon, Wear-dale, and different Swedish brands, mixed in propor-tions known only to Perry & Sons, to make soft and chilled rolls, for rolling plates, sheets, armour plates, &c., &c.; for the manufacture of which this firm is justly celebrated.

Here some of the rolls used at the Earl of Dudley's, W. Barrows & Sons, the Barrow Steel Company, Robert Heath & Sons, the Blochairn in Scotland, and most other great concerns, are made.

We know of no concern which has done more to raise the character for engineering skill in the Black Country than Thomas Ping & Sons. This firm often take contracts for complete Iron works for foreign countries. They have a good name for blast engines; all the blast engines at the Steel Company's works at Barrow were built by this firm—safes of a very high class are made here, and other specialities out of the range of the fitting business.

THE ALBION IRON WORKS, WEST BROMWICH

Represented by the annexed engraving, were erected at great cost thirty years since, in the most substantial manner, by the late Walter Williams, Esq.

The plan, general arrangements, and architecture of the Albion Works reflect credit on the judgment of the late Mr. Walter Williams, who took pride in the works, and always sent a quality of Iron into the market from these works which established their first-class brand throughout the country. The late proprietor was the eldest son of the first Mr. Philip Williams, of Wednesbury Oak, the founder of the well known 'Mitre' brand of Iron.

These extensive works are now carried on by several firms. The Albion Sheet Iron Company and the Britannia Iron Company (the proprietary being almost identical) have here two forges, consisting of nineteen puddling furnaces, two sheet mills, and a bar mill, where are made sheets of very superior quality of the well-known 'Trident' brand, and bars—small rounds and squares—with various kinds of fancy Iron, marked 'Britannia Iron Co. ❦.'

Their sheets, bars, and specialities in Iron, take a high position in the market. The firms consist of Titus Greenway, William Dangerfield, Henry Lewis, Edward W. Lewis (the Albion Sheet Iron Company), and of Henry Lewis, Edward Lewis, and William P.

BRITANNIA IRON Cᵒʸˢ ALBION WORKS, WEST BROMWICH.

Greenway (th
years Mr. T
of sheet mill
Delaston G
quarter of a c
Walker, of th
celebrated Gc

The Messrs
Wednesbury
Sons, where t
and Mr. W.
rience under
Mills, of the
laston Gree
fore a practi

TANGYE B

THESE engi
machinery
A hundred
metalliferc
mineral lo
sands of m
or coal m
hundreds c
treasure fr
the earth u
counties th
on a farm

Greenway (the Britannia Iron Company). For many years Mr. Titus Greenway held the responsible position of sheet-mill manager to the late Mr. Samuel Mills, of Darlaston Green. Mr. Dangerfield was, for over a quarter of a century, mill manager to Messrs. G. & E. Walker, of the Gospel Oak Iron Works, makers of the celebrated GO ᵼ sheet Iron.

The Messrs. Lewis were instructed in Iron making at Wednesbury Oak, under Messrs. Philip Williams & Sons, where the famous ' Mitre ' brand of Iron is made ; and Mr. W. P. Greenway has had considerable experience under our own relative, the late worthy Samuel Mills, of the well-known Bills and Mills firm of Dar laston Green. Every member of these firms is therefore a practical Iron maker

TANGYE BROTHERS AND HOLMAN'S DIRECT-ACTING STEAM PUMPING ENGINE.

THESE engines have created a great revolution in the machinery now in use for hydraulic purposes in mines. A hundred years since water in coal, Iron, and all metalliferous mines, was the great enemy alike of the mineral lord and the sturdy adventurer; hence thousands of mines were abandoned the moment the lode or coal measure dipped into the water-pond, and hundreds of thousand pounds' worth of subterranean treasure from this cause was left by the *old men* in the earth unutilised. Hence it often occurs in mining counties that an old pit-shaft is discovered by accident on a farm now devoted to agriculture, where every

These wo...
made their o...
worked s...
of water at...
16 tons of...

At a...
was i...
lead in t...
on what is...
been in ro...
of these...
The founda...
quently...
pounds...
of the...
taken as i...
the contra...
smaller an...
decessor...
machinery...
valves, t...
condens...
it rende...

vestige of pits or mining appliances are removed by the hand of time and agricultural operations. *Tempus edax omnium rerum*. About the year 1782, the Carron Company in Scotland erected a water-engine which, for its dimensions and capabilities, was looked upon as one of the wonders of Scotland. Sir J. Sinclair, who wrote in 1792, speaking of this engine, says :—

'These works consisted of five blast furnaces, and made their own fire-bricks, had a Water Engine, that worked seven strokes a minute, raised 3,500 gallons of water at one stroke. The same engine consumed 16 tons of coal in 24 hours.'

At a more recent date more powerful machinery was invented, and our friends in Cornwall took the lead in this respect, and constructed steam-engines on what is called the 'Cornish plan,' which have since been in vogue at all our metalliferous mines. The cost of these engines, however, is a bar to their adoption. The foundations, house, and other appliances frequently costs ten, fifteen, and even twenty thousand pounds, much in this respect depending on the depth of the mine to be drained. These remarks must not be taken as inveighing against the Cornish engine. On the contrary, we say they do more service with a smaller amount of fuel by far than any of their predecessors, but the cost of foundations, the complicated machinery, the beam, the massive centre gudgeon, the valves, the levers, the hand gears, the air-pump and condenser ; in fact, the whole machine, for the service it renders, cannot be compared with the unique and

scientific apparatus in the engraving annexed, invented
by Mr. A. S. Cameron, of New York.   Already 3,500
of these engines are fixed and at work, all made at

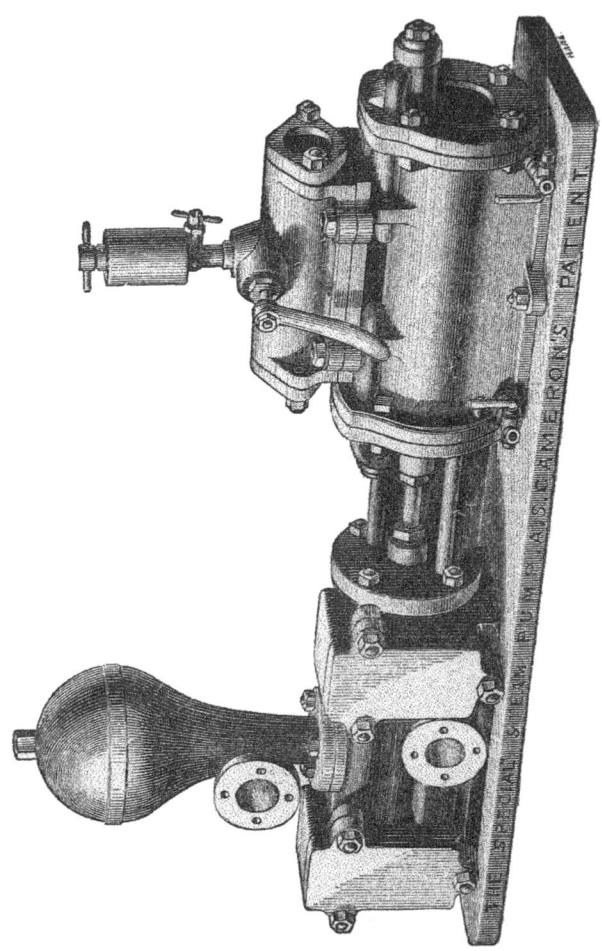

the great Cornwall works of Tangye, Bros., at Soho
and Birmingham.

It will be observed by the engraving that there is

nothing to be seen moving but the piston rod in its constant reciprocations, the steam cylinder to the right and the pump to the left. The power exerted by the steam on the piston is expended in the pumps through the medium of the piston rod itself, the reversing valves are operated by the piston, all is closed up, nothing exposed to view, and the speed of the engine is retarded or accelerated *ad libitum* by the regulation of a simple supply valve.

The great advantage of these pumps is the facility with which they can be laid down; like the horizontal engine they are fixed with very little cost, they rarely get out of order, the seats of the valves being made of vulcanized india-rubber or something of this kind. A gentle sounding click is heard, which has no ring of metal, but rather gives the idea that everything is tight and sound like the air pump bucket of a condensing engine when it has been recently packed with a good white rope. These engines are now, we understand, fixed in the nethermost depths of the mine, and lift the water to the top; this plan answers well, preventing the strain and continual disturbance of the water trees where the old engines are used. The steam may be generated in the pit or at grass at pleasure; of course, if the latter, steam pipes are fixed to descend the shafts to the engine. They can be fixed of any size required by the work to be done, and the small cost when compared with the old engines is marvellous and astounding, and no doubt they are rapidly coming into universal use in all parts of the world. The works of this great firm are situated very near to the old Soho Works, near Birmingham. They have likewise another large factory in Birmingham. In looking over the Soho Works the other day, we were pleased more than we can express. They are of immense size, and employ twelve hundred men; and the wholesale way in which the work is got through is marvellous. We should think there were fifty or sixty of these engines of different sizes in course of erection. We frequently saw two or three cylinders being bored by the same machine at the same time. The slotting, drilling, boring, punching, turning, planing, forming, and forging machines, with all the men actively at work, was gratifying to witness. Here hydraulic power is used for cutting, punching, forging, &c.

The foundries and casting shops in brass and Iron, the smithies and chain-making shops, are all abreast, if not in advance, of the day in which we live. The offices, and even the warehouses for exporting the produce of this wonderful hive of industry, were admirable in the highest degree. There is likewise here a spacious dining-hall for the men, with stoves and cooking apparatus under proper management, where the artizans may have tea and coffee for breakfast and their chop or steak cooked for dinner, and no doubt, although we did not enquire and cannot speak positively, a library is kept and a sick club under the management of the able gentlemen who established, and with such spirit and regularity carry on, this great concern of Tangye, Bros. At these works, besides the Special direct-acting Steam Pumping Engines referred to above, are made Hydraulic and Screw Lifting Jacks, Cranes, Differential and ordinary Pulley Blocks, and a very large number of their newly designed Horizontal High Pressure Steam Engines. Also Ramps for speedily re-railing Railway Carriages in the event of their getting off the line. Hydraulic and Screw Presses for every variety of purpose. Hydraulic Testing Machines, Morton's Patent Self-sealing Retort Lids with Holman's Patent Fastenings, Holman's Patent Purifier Lifts and Double-acting Pumps, Oesterkamp's Patent Rock Borer, and a great variety of special articles.

The Messrs. Tangye, Brothers, we believe, have a Depôt at Newcastle. The London business offices are at 10, Laurence Pountney Lane, E.C., under the direction of Mr. Holman, the managing partner in London.

PARTICULARS OF MANUFACTURERS AND THEIR QUALITIES.

N. Hingley and Sons: furnaces, 4, 3 in blast; puddling furnaces, 42. Make best cable iron. Old Hill and Nertherton furnaces. Brand: N.H. & S. ✿.

Pontnewynyod Iron Works, Pontypool: pudd. furnaces, 26, and building more. Make wire rods, tin bars, and sheet iron. Iron quite first class. No B. furnaces.

Wilden Iron Works, near Stourport: proprietors, J. P. and W. Baldwin; one of the oldest works in Worcester: sheet iron and tin plates, both of the very best quality. Charcoal, sheet iron, and tin plates. Brands: all sheet iron branded 'BALDWIN-WILDEN,' and marked with distinguishing quality, 'B,' 'B B,' 'B B B,' 'B. CHAR'L,' 'E.B. CHAR'L.' Tin-plates for deep-stamping — branded $\frac{\text{'E.P.} \times \text{W.B.,}}{\text{W.H.}}$ BEST CHAR-COAL TIN.' Tin-plates for general working — branded 'WILDEN-CHARCOAL,' 'UNICORN-CHARCOAL TIN,' 'ARLEY ✿ TIN.' Coke tin-plates — branded 'STOUR COKE TIN.' Tin-plates specially prepared for gas-meter purposes. Our sheet iron is not only marked 'B,' 'B B,' &c., but the brand 'BALDWIN-WILDEN' is on the bands of each bundle.

Withymoor Furnaces: 2 in B. Make best hydrate iron. W. Dawes and Sons, proprietors.

Henry Hall, Old Level Works, Brierley Hill: 18 puddling furnaces. This iron is very good. Best and best best bars— rounds, $\frac{3}{16}$ in. to 4 in. diameter; squares, $\frac{1}{4}$ in. to $3\frac{1}{2}$ in. square; flats, $\frac{1}{2}$ in. to 6 in. wide; hoops and slit rods; colliery rails; T angle and fancy iron. Brand ✿ LEVEL ✿, HALL'S B B ✿.

The Darlaston Steel and Iron Company, established 1799: 42 puddling furnaces, 3 B. F., all in B. Brand. The undermentioned Brands are the Trade Marks and Brands of the Company, and are stamped upon iron and steel at their Works, and are used to distinguish the various qualities of iron and steel made at the Works, all the very best :—

TRADE MARK.

No. 1. 'CHAMPION.' No. 2. 'L ✿ DAR-
LASTON.' No. 3. 'SAMUEL MILLS.'
No. 4. 'G. F. ✿ G. F.' No. 5. 'LLOYD'S
DARLASTON.' No. 6. 'LLOYD'S
CHARCOAL.' No. 7. 'L ✿ STEEL.'

No. 8. '⟨◇ I⟩ STEEL.'

CHAMPION.

Gold, Brothers, Sudely Furnaces, Forest of
Dean: 2 B. F., 1 in B.

The Ashton Vale Iron Company, Bristol. Very good iron. 1
B. F., 1 in B.

Gjers, Mills, and Company, Ayresome Iron Works, Middlesborough : 4 B. F., 4 in B.

Appleby and Company, Renshaw Iron Works, Chesterfield :
4 B. F., 3 in B. Established 1785. Large iron founders.

The Chillington Iron Company : 95 P. F., 6 B. F. Brand, CC✿,
LB ✿, Chillington.

Tredegar Iron Company: 80 P. F., 9 B. F., 7 in B. Rails and
Fish Plates. Brand T. I. CO.

Fletcher, Solly, and Company, Willenhall : 3 B. F., 3 in B. Very
good iron.

Lloyds and Company, Middlesborough : 6 B. F., 6 in B. Brand,
'Linthorpe.'

Bilston Brook Furnaces : 3. B. F., 2 in B. Cinder foundry and
forge iron.

M. S. Goddard and Sons, Lane End Iron Works : 2 B. F., 2 in
B. About to erect 1 new furnace.

Messrs. Round : 2 B. F., 2 in B. Best mine iron.

Roberts and Company, Tipton Green Furnaces : 4 B. F., 4 in B.
Good melting iron for making bedsteads and forge iron.

Watson, Kepling, and Company, Seaham : 1 B. F., 1 in B. Erecting one new furnace, 1871. Make first-class Bessemer iron.

North of England Iron and Coal Company (Limited), Carlton
Works, Stockton-on-Tees : 3 B. F., 3 in B. 8 patent Danks's
puddling furnaces.

Thomas Webb and Sons, Bretwell Hall and Bretwell Lane Iron
Works, Stourbridge : 21 puddling furnaces. Hoops, rods, and
angle iron. Brand.

Sevann, Coates, and Co. : 4 B. F., all in blast. Brand, Cargo
Fleet.

The Wingerworth Iron Company, near Chesterfield: 3 B. F. all in B.

Lumphinnans: 2 B. F., 1 in B.

Frodington Works: 4 B. F., 4 in B., 1864.

The Weardale Iron Company, Towlaw Iron Works: 4 B. F., 2 in B. 1 undergoing repair. This is very superior iron, quite the best made in this district. No. 3 Pig being worth 8l. per ton to-day.

Messrs. Molineux, Bull's Bridge Iron Works, Morley. Make best thin sheets. Brand, a bull's head.

Messrs. Fowler and Company, Barbor's Field Iron Works, Bilston: 2 B. F., 1 in B. Very best forge mine iron.

Mr. John Marshall, Monway Iron Works, Wednesbury: 10 P. F. Bars, sheets, all best and best best charcoal, and extra best charcoal for the government. Price to-day, from 33l. 10s. per ton to 5d. per lb. Mr. Henry Mills, of the Pleck Iron Works makes the same qualities and gets the same price.

Hugh Martin and Sons, Coatbridge Iron Works: 7 P. F. First class rivet iron.

Brymbo Iron Company, Wrexham: 2 B. F., 2 in B. 1 new one building; established 1797.

Lee and Bolton, Hyde Iron Works, Stourbridge: 20 P. F. 2 Siemens' furnaces at work here; brand of iron first class. These are works that first rolled slit rods, which were introduced by the ingenious Foley, who obtained admission into the Russian Iron Works by the charming strains of his violin. See account of his journey in the *Guide*. Brand, ✿ L & B. This is perhaps the oldest works in Worcestershire, always makes first class iron, and has the Honourable Foley's first made rods here. Lee and Bolton continue this trade. The works are situated in a lovely spot and charming country. See Engraving. They likewise make plating, bars, wire rods, best bars, and best sheets, and obtain the highest price.

Norton Iron Company (Limited): 5 B. F., 4 in B. 1 of these working on Norwegian Titanic ore making Bessemer iron, and for this 8l. 5s. per ton is now obtained.

Robert Crawshay, Cyfartha Iron Works, Merthyr Tydvil, one of the oldest works in England, established 100 years : 11 B. F. All on strike. Brand, W.C. Make bars and rails at the works. 72 P. F.

W. Barrows and Sons, Bloomfield Factory, and Tipton Works : 100 P. F. Makes 60,000 tons a year of the very best Staffordshire iron. First quality, has a world-wide fame. Brand, $\overset{\heartsuit}{\text{B.B.II.}}$ generally fetches 1*l.* per ton more than all other Staffordshire makes, SC ✿ and the Earl of Dudley's excepted.

Mr. John Spencer, Phœnix Iron Works, Coatbridge, Scotland : 22 P. F. Bars, angles, and boiler-plates, high class quality. Brand, Phœnix.

Snedshill Iron Company, Snedshill, Salop : make bars, plates, sheets, and wire rods. Quality of all the iron made here is first class, being entirely made from Earl Granville's best Lilleshall pigs. Brand, $\dfrac{\text{H. S ✿ BEST}}{\text{SNEDSHILL.}}$ Mr. Thomas Horton, Prior's Lee Hall, and his brother, Mr. Samuel Horton, are partners in this concern. They deservedly stand very high in the market for all kinds of iron they make, and have a number of charcoal fires, and produce charcoal wire rods in perfection.

Britannia Iron Company, at the Albion Works, West Bromwich, has 10 puddling furnaces, where bars, small angles, rounds, squares, half rounds, and all kinds of fancy iron are made. The quality is of the highest class. The brand, BRITANNIA IRON CO. ✿. The iron made here fetches high prices.

Albion Sheet Iron Company, West Bromwich, have 10 puddling furnaces, and 2 sheet mills and a bar mill, where the famous ' Trident ' sheets are made which fetch such high prices. This and the firm above are noted practical men in iron making.

George Adams and Son, of the ' Mars ' Priestfields Iron Works, Wolverhampton, have 12 P. F. and 3 mills. Make bars, sheets, and hoops of the very highest class. Mr. George Adams being a very able and experienced manager in this capacity, he contributed by his able management to the fame of the brand of Rose, Higgins, and Rose's iron, of Bradley Field, and Wright and North's, of Wolverhampton, to whom he was chief manager for years. His own brand of iron now stands quite as high as either ; a plate of these works will be found in the *Guide.* Brand, ' Mars.' All best iron made here, best, double best, &c., &c.

William Millington and Company purchase all their pigs, have no blast furnaces. They have 20 puddling furnaces, and 4 mills and forges, situated at Summer Hill, Tipton, railway station near to the works. These works have been established upwards of 40 years, and stand well in the market for the quality of their boiler plates: Brand, W M Co. surmounted by a dove, or S H Crown. Besides boiler plates, they make angles, plating bars, rivet iron, half rounds, ovals, small rounds, and squares. All their qualities of iron are good, and may be shipped with confidence. Their make of finished iron is 7,000 tons per annum. Their price to-day is 15l. 10s. for bars, and 19l. 10s. for plates. In the Paris Exhibition their iron was 'honourably mentioned.'

The District Iron and Steel Company is situated at Smethwick, 300 yards from the Stour Valley railway station. Are very good works, capable of rolling most kinds of iron, were built 50 years ago, have 4 mills, 3 forges, and 20 puddling furnaces. The proprietors of this extensive concern have likewise other works at Church Lane, Tipton, with 10 puddling furnaces and 2 mills. There are few works in the Black Country laid down so well. At the District, the arrangements are good, near to railway, and hangs over the canal like Capri in the Bay of Naples, The engines are all up to their work, and sheet iron, hoops, bars, angles, small rounds, and squares are made here, and at Church Lane, of very good quality. Indeed, we have known these works 40 years, and they have always had the character for making good iron. The proprietors are Messrs. Broughton, Keen, and Hawkins, well known in the trade for a long period.

The Thornaby Works are situated at Stockton, carried on by Messrs. William and Thomas Whitwell. Have been in business 11 years. There are 34 puddling furnaces and 3 mills. 'They manufacture rounds, squares, flats, half rounds, convex angles, Tees rails, channel, and cable iron, and turn out 13,000 tons of finished iron per annum. They have 3 blast furnaces, all in blast, and make 48,000 of pig iron per annum. Now engaged in erecting 2 new ones. Brand, a Mallard Lion, W W & Co., Cleveland. Pig iron brand, THORNABY. Character of iron, good.

The following is a List of all the London Iron Merchants and Exporters of Iron, coals, and all kinds of hardwares, describing the particular Ports to which they trade ; likewise Merchants and Exporters of copper, tin, lead, spelter, bullion, &c. (See our remarks at the end of the List.)

W. Warehousemen and Shippers.   M. Manufacturers and Shippers.

Warehousemen and Shippers.	Ports.	Manufactures.
Abrahams, Hyam, A., 87, Houndsditch	Various . . .	Birmingham and Sheffield goods
M. Adam's Patent Small Arms Co., 391, Strand	Various .	Revolvers and fire-arms
Adamson & Ronaldson, 1, Leadenhall Street	Colonies, Natal, Australia, Cape, China, and East and West Indies	Metals, files, saddlery, hardware, gunpowder, woolpack, and general
Adutt, Finzi & Co., 24, Mark Lane	Levantine Ports, Baltic, and United States	Colonial produce, drugs, metals, fruits, Manchester goods, machinery, &c.
Agricultural and General Machinery Agency (Lim.), (Wm. Smith, manager), 19, Salisbury Street, Strand	Various . . .	Machinery
Aldridge, Joseph, F., & Co., 4, East India Avenue	India and China .	Apothecaries' wares, apparel, beer, ales, hardware, cottons, earthenware, and machinery
Alexander, Fletcher & Co., 10, King's Arms Yard	Calcutta, Madras, Bombay, and China	Linens, machinery, hardware, and general
M. Alfred & Son, 54, Moorgate Street	India, Australia, &c. .	Fishing-tackle, &c.
Allen & King, 7, Dowgate Hill	Continent and various	Iron and machinery
Allen Brothers & Co., Albion Place, London Wall	East Indies . . .	Drapery, hardware, machinery, cottons, perfumery, jewellery, &c.
Allport, Douglas, 9, Fenchurch Buildings	West Indies and Natal	Cottons, linens, woollens, hardware, and general
M. Allwood & Sons, 11, London Wall	Various . . .	Needles
W. Almond, W. J., 174, Aldersgate Street	Various . . .	Shoe threads, nails, elastics, needles, &c.

Warehousemen and Shippers.	Ports.	Manufactures.
Alpe, J. P., & Co., 149, Fenchurch Street	East and West Indies, and others	Manchester goods, iron, hardware, medicines, apparel, machinery, and general
Abrutz, R., 96 & 97, Ethelburga House	Various . . .	Iron
Anderson, T., 27, Leadenhall Street	Melbourne and Auckland	Timber, provisions, hardware, &c.
Angier, S. H., Moore & Co., 72, Cornhill	Riga and Russian Ports	Woollens, cottons, and earthenware
Ashby, Morris, 17, Laurence Pountney Lane	Various . . .	Zinc, oxide, and sheet zinc
Appleyard, C. H., & Co., 10, Camomile Street	Various . . .	Hardware
M. Arnold & Sons, 35 & 36, West Smithfield	All over the world .	Surgical and retinary instruments, cutlery, &c.
Aria & Co., 70 & 71, Bishopsgate Street Within	Jamaica, Honduras, and Calcutta	Manchester goods, hardware, &c.
Ashford & Brookes, 29, Great St. Helen's	Sydney, Adelaide, and Melbourne	Hardware, earthenware, glass, and general
Ashton & Co., Hatton Court, Threadneedle Street	East and West Indies, China, &c.	Manchester goods, hardware, electro-plate, cutlery, iron, machinery, beer, spirits, wine, and general
Assam Company, The (H. W. Wimshurstf, secretary), 2, East India Avenue	Calcutta . . .	Metals, tools, provisions, blankets, &c.
Atkins, Charles, & Co., 1, Water Lane	Continent . . .	Manchester and Yorkshire goods, machinery, and general
Austin Brothers, 81, Gracechurch Street	Cape Town and Cape of Good Hope	Coals, metals, hardware, &c. (Ship brokers)
Ayling, W.W., & Co., 24, New City Chambers, Bishopsgate Street	Australia, New Zealand, East and West Indies, Brazil, and Continent	Apothecaries' wares, drapery, haberdashery, machinery, &c.
Azzoni, Francesco, 25, Old Broad Street	Continent and America	Iron and manufactured iron goods
Bailey, Pegg & Co., 81, Bankside, s.e.	East Indies, China, Japan, and various	Iron, metals, arms, &c.
M. Bailey, Wm., & Son, 2 & 3, Abchurch Yard	East India, China, Australia, South America, &c.	Chemicals and articles for telegraphy
Bailey, W., & Sons, 71, Gracechurch Street	Cape, West Indies, South America, and Brazils	Ironmongery and implements
Bain, Stead & Elford, Golden Heart Wharf, Dowgate	East Indies . . .	Metals and general

Warehousemen and Shippers.	Ports.	Manufactures.
Baker & Oliphant, Walbrook	East Indies. . .	Surgical instruments and general
W. Baker, R., & Co., 309, Oxford Street	Australia, &c. . .	Agricultural implements and mangles
Balleras & Liggins, 11, Leadenhall Street	Spain and various .	Iron, coals, railway plant, and general
Banks, Wm., 32, Lombard Street	Spain and Continent .	Coal, iron, machinery, and rails
Barber, James, Son & Co., 136, Leadenhall Street	Cape, East Indies, and Australia . .	Hardware, colours metals, leather, tools, soft goods, and general merchandise
W. Barnes, Fred., & Co., 109, Fenchurch Street; Birmingham and Sheffield	Australia, East Indies, Coast of Africa, South America, Brazils, &c.	Hardware, arms, ammunition, tools, general ironmongery, &c.
M. Barrows & Stewart, 97, Cannon Street	Australia and New Zealand	Steam engines and mortar mills
Barter, W., & Co., 38, Gracechurch Street, and Lloyd's	Mediterranean, &c. .	Coals
Bartrum, Pretyman & Mumford, 168, Upper Thames Street	Australia, Cape, &c. .	Iron and copper
Basden, Townshend & Co., 11, Great St. Helen's, and Lloyd's	Russia .	Machinery and iron
Bateman, W., & Sons, 10, Clement's Lane; and Baltic	Tasmania and New Zealand	Apparel, haberdashery, hardware, drapery, &c.
Bayley, J. A., 17, Gracechurch Street	Italy and West Indies	Hardware, &c.
Belaieff, A. P., 10, East India Chambers	Russia and North of Europe	Colonial produce and machinery
Bell, Brandenburg & Co., 2, Billiter Square	East Indies . . .	Cottons, hardware, machinery, &c.
M. Bellamy & Smith, 5, Abchurch Yard; also Sheffield	Australia, Spain, and Germany	Hardware, files, &c.
Benohr, Henry, & Co., Finsbury Place	Italy . . .	Manchester goods, hardware, &c.
Berger, Leo, 3, Philpot Lane	Hamburg, Reval, Trieste, &c.	Metals
Beriro & Co., 5, Idol Lane	Spain, Portugal, and Morocco	Toys, cottons, linens, iron, colonial produce, &c.
M. Bessemer Brothers, East Greenwich	Various . . .	Iron and steel manufactured goods
M. Binks Brothers, Stafford Street, Millwall	South America, Australia, &c.	Galvanized iron and zinc wire
Bird, W., & Co., 2, Laurence Pountney Hill	Continent, British, Colonies, and America	Iron, machinery, metals, and contractor's plants
Birdseye, H., & Co., 5, St. Benet's Place, Gracechurch Street	East Indies, China, Cape, and New Zealand	Straw, hats, caps, wines, mineral waters, Manchester goods, hardware, &c.

Warehousemen and Shippers.	Ports.	Manufactures.
Bischoffshiem & Gold-schmidt, Founders Court, Lothbury	Various . . .	Bullion
M. Bishop, E., & Co., 6, Walbrook; and Birmingham	Spain, Australia, &c. .	Hardware
Blackwood, Conor & Co., 27, Mincing Lane	East and West Indies, Australia, and Mexico	Cottons, linens, glass, hardware, earthen-ware, coals, bricks, sundries
Blair, J. F., 21, Abing-don Street, West-minster	Various . . .	Machinery
Blakemore, V. & R., 46, Leadenhall Street; and Birmingham	Various . .	Hardware and arms
Blannd, John, 123, Feu-church Street	Gibraltar . . .	Coal
Blandy, Charles R., 25, Crutchedfriars	Madeira . .	Coals and general
M. Blews, Wm., & Sons, 38, West Smithfield; and Birmingham	Mexico, Spain, &c. .	Church bells and brass founders
Blundell, T., & Co., 52, Gracechurch Street	India, &c. . . .	Brass and copper
Blyth, Greene & Co., 3, King William Street, E.C.	Mauritius and West Indies	Iron, coal, and British manufactured goods. (Shipowners)
Boddington & Co., 9, St. Helen's Place	West Indies, Brazils, and South America	Cottons, linens, hardware, provisions, glass, &c.
Bolitho, T. & W., 39, Lombard Street	Continent and all parts	Tin
Bordiu, Fabris & Co., 44, Coleman Street	Continent . . .	Bullion and general
Born & Co., 13, Ber-ner's, Oxford Street	Hamburg, Stettin, and Rotterdam	Hardware, colonial pro-duce, carpets, bricks, drain pipes, and agri-cultural implements
Borneo Company (Lim.) (J. Harvey and W. Martin, managers), 7, Mincing Lane	East Indies and China	Manchester goods, me-tals, machinery, and general
Boucher, Guy & Co.,123, Leadenhall Street -	East Indies, &c. . .	Porcelain, glass, and earthenware
Boustead, E., 5, New-man's Court, Cornhill	Singapore and Penang .	Cotton goods, hardware, earthenware, and gene-ral merchandise
Bovet, F. & A., 150, Leadenhall Street	China	Jewellery, cottons, wool-lens, and metals
Bowring, Arundel, & Co., 12, Fenchurch Street	East Indies and China	Hardware, machinery, apparel, beer, ale, cot-tons, earthenware
M. Braby, Fred., & Co. (Lim.), 17, Grace-church Street	Various . . .	Galvanized corrugated iron and zinc

Warehousemen and Shippers.	Ports.	Manufactures.
M. Bradford, Thomas, & Co., 63, Fleet Street; also Manchester	Various . . .	Washing, wringing, and mangling machines
Bradshaw. Henry, 129, Lower Thames Street	France, Continent, and Mediterranean	Coals and iron
Brand, Jas., 109, Fenchurch Street	West Indies and United States	Cottons, woollens, glassware, hardware, linens, machinery, tinplate, furs, skins
Brandon, Jonathan (agent for F. M. Brandon), 12, Fenchurch Street	Rio de Janeiro . .	Cottons, linens, iron, machinery, fancy dress goods, and every description of small ware
Bravo, Joseph, & Co., 3 & 4, Great Winchester Street Buildings	West Indies . .	Cotton goods, provisions, arms, ammunition, and general
W. Breillat, Joseph, 27, Blackman Street	India and Colonies .	Glass, china, earthenware, &c.
British Honduras Company (Lim.), 2, Great St. Helen's	Belize and British Honduras	Cottons, linens, hardware, metals, machinery, woollens, paints, oils, &c.
M. Brook, Henry, & Co., 10, Featherstone's Buildings	Amsterdam, India, and China	Garden engines and India-rubber goods of every description
W. Brooke, R. J., 9, Houndsditch	Various .	Hardware
Brookes, Robinson, & Co., 73 & 74, Ethelburga House; 171, Bishopsgate Street Within	Colonies .	Window-glass, hardware, &c.
M. Brown Brothers, 165, Piccadilly	India, China, and various	Portable iron furniture, &c.
M. Brown & Green, 72, Bishopsgate Street Within	Australia, India, and Cape of Good Hope	Kitchen ranges, portable cooking stoves, velocipedes, &c.
Brown & Mainnett, 26, New City Chambers	Spain and West coast of Africa	Iron, arms, and general hardware
M. Brown, Sir John, & Co. (Lim.), 10, John Street, Adelphi; also Sheffield	Various . .	Iron and steel manufactured goods
M. Brown. Lenox & Co., 8, Billiter Square	East Indies, Colonies, United States, China, &c.	Anchors, cables, patent moorings, and machinery
Brown, Walter H., & Co., 11, Billiter Square	Various .	Coals, tar products, pitch, creosote, naptha, petroleum, chemicals, drysalteries, and colonial produce
M. Brown, Westhead, Moore & Co., Holborn Viaduct	Various . . .	Earthenware and china

Warehousemen and Shippers.	Ports.	Manufactures.
Bruce, G. G., 60, Gracechurch Street	West coast of Africa, Canaries, Madeira, St. Michael's, Teneriffe, Mediterranean, South of Europe, &c.	Cottons, provisions, arms, ammunition, and general
Bryan Brothers, 36, Crutchedfriars	Colonies, Mediterranean, and various	Cottons, linens, machinery, and colonial produce
Bryant, George, Lime Street Chambers	Australia, New Zealand, East Indies, &c.	Iron, tin plates, metals, and general
Budd, E. L.. & Co., 8, Moorgate Street	East Indies, China, Continent, &c.	Metals and general
Budden, Jennings & Co., 48, Fenchurch Street	East Indies, China, Australia, Cape, and other	Metals, Manchester, Birmingham, Sheffield goods, machinery, &c.
W. Burgess & Key, Holborn Viaduct	Colonies and India	Reaping and mowing machines, cotton gins, &c.
Burnell, Martin & Co., 5 & 6, Great Winchester Street Buildings	Algoa Bay, Natal, Mauritius, and Yokohama	Apparel, hardware, drapery, woollens, blankets, and general
Burnes, J., & Son, 138, Leadenhall Street	Aden, Cape and various	Steam and other coals
Burnuss, James W., Dunster House, Dunster Court	Shanghai, &c.	Fibre, minerals, and coals
Burr, D., MacGibbon & Co., 204, Up. Thames Street; also Falkirk, N.B.	Various	Ironware and milled plates
Byrne, R., & Co, 137, Fenchurch Street and Cardiff	Mediterranean, Black Sea, Baltic, &c.	Coals and iron
Calway, Bartholomew, 58, King William Street	Adelaide	Hardware
Campbell, John, & Co., 15, Austin Friars	East Indies .	Hardware, machinery, cottons, woollens, and piece goods
Carnegie, A., 16, Bishopsgate Street Within	Galatz, Ibrail, Malta, Constantinople, Syria, and Odessa	Hardware, earthenware, ironware, and general
Carr, C., & Co., 14, Bishopsgate Street Within	Mediterranean, Continental Ports, America, China, and Japan	Manufactured goods, machinery, &c.
Chalmers, Guthrie & Co., 9, Idol Lane, Tower Street	Cape, Mauritius, Indies, Central and South America, and China	Colonial produce, wines, beer, cotton goods, arms, ammunition, and general merchandise
M. Chatwood, Samuel, 12, Cannon Street	Various	Fire-proof safes
Child, Hornby & Co., 27, Lombard Street	West Coast of Africa, India, Egypt, and North and South America	Cotton manufactures, hardware, coals, &c.

Warehousemen and Shippers.	Ports.	Manufactures.
Chinnery & Johnson, 67½, Lower Thames Street, and Folkstone	Paris, Boulogne, Dieppe, and Dunkirk	Leather, cottons, silks, woollens, haberdashery, linens, steel, and iron
M. Chisholm, J., Son, & Co., 44, Mark Lane	Various . . .	Agricultural implements and machinery
Chollett, M., & Co., 2, Ingram Court	Trieste, Adriatic, and various	Manchester, Sheffield, and Birmingham manufactures, and preserved provisions
M. Chubb & Son, 37, St. Paul's Churchyard	Various . . .	Iron safes and locks
M. Churchill, Chas., 28, Wilson Street	Australia and various .	American hardware
Clauson, Charles, 157, Fenchurch Street	Mediterranean .	Metals, &c.
Clay, Cooper & Co., 3, Adelaide Place, London Bridge	New Zealand, Alexandria, Black Sea, and various	Metals, hardware, and general merchandise
Cobb, A. B., & Co., 34, Great St. Helen's	Australia . . .	Drapery, haberdashery, machinery, &c.
Cohen, Aaron, & Co., 161, Great Dover Street	America and various	Rags, metals, and general
Cohen, D., 41, Sun Street, Finsbury	Australia, Continent, &c.	Metals
Cohen, Mylius, 57, Gracechurch Street	Continent and various .	Metals, &c.
Coles, Brown, Andrews & Co., Levant House, St. Helen's Place, and 19, St. Helen's Place	China, Japan, and India	Metals, cottons, linens, woollens, piece goods, and general
Colla, D. M., 4, Cullum Street	Greece . .	Hardware, glass, brushware, woollens, and general
Colonial Company (Lim.), 16, Leadenhall Street	West Indies . .	Apparel, saddlery, hardware, and general
M. Colt's Fire Arms Company, 14, Pall Mall	India and Colonies .	Revolvers, &c.
Colville, E., 9, Fenchurch Buildings	West Indies . .	Hardware and estate supplies
Compton & Read, 8, Moorgate Street	East Indies, New Zealand, China, and Australia	Apparel, cottons, machinery, and hardware
Compton, H., & Co., 148, Fenchurch Street	India, Africa, and all other parts	Pewterware, tin-foil, and tea-lead
Compton, Wohlgemuth & Hardness, 4, Coal Exchange, Lower Thames Street	Various . . .	Coals
M. Copeland, W. T., & Sons, 160, New Broad S reet	East Indies and Colonies	Glass, china, and earthenware

Warehousemen and Shippers.	Ports.	Manufactures.
Corcos, Abraham, 45, Houndsditch	Mogadore . . .	Manchester goods, cloth, earthenware, and china
Cory Bros. & Co., 150, Leadenhall Street	Lisbon, Malta, Port Said, &c.	Coals
Cotesworth & Powell, 16½, St. Helen's Place	Australia, China, South America, and East Indies	Cottons, iron, general and British manufactured goods
Coulon, Berthoud & Co., 21, Threadneedle Street	Continent and various	Oils, colonial produce, and metals
Coulthard & Co., 12, Abchurch Yard	East Indies . . .	Metals, hardware, &c.
Cowderoy & Rainbow, 5, New London Street	China and East .	Apparel, machinery, cotton, and general
Cox, Rowland, 3 and 4, Great Winchester Street Buildings	Brazils . .	Hardware and general
M. Crane, T. H., 3, Royal Exchange	India, Australia, &c. .	Guns
Crawford, Colvin & Co., 71, Old Broad Street	East Indies and China	Hardware, metals, Manchester goods, and general
M. Croggon & Co., Albion Wharf, 10, Upper Thames Street	Various . . .	Patent felt, zinc, galvanized iron, &c.
Cuadra, B. de, 43, Lime Street	Australia, New Zealand, East Indies, and China	Wearing apparel, apothecaries' ware, drapery, haberdashery, earthenware, and hardware
Cunningham & Co., 480, Oxford Street	All parts of the world	Sewing, sawing, metal-cutting machines, &c.
Cunningham, R. S., 158, Leadenhall Street	India, Australia, Cape, and South America	Metals, hardware, India rubber, patent paints, and general
Curtis, J. L., Palmerston Buildings	New Zealand and Australia	Machinery, hardware, and general merchandise
Culbill, Son & De Lungo, 103, Cannon Street	East Indies, &c. . .	Metals, railway iron, &c.
Da Silva & Co., 26, Bride Lane, Fleet Street	Various . . .	Hardware, medicines, and general merchandise
Dadabhai, Naoroji & Co., 32, Great St. Helen's	East Indies . . .	Machinery, piece goods, yarns, fancy goods, &c.
Daniell & Co., 25, Old Broad Street	East Indies . .	Cottons, woollens, metals, and general
Daniels, F., 2, Riches Court	United States . .	Railway materials, hardware, and general goods
Dart, J. H., & Son, 3 and 4, Great Winchester Street Buildings	Spain and Portugal .	British manufactured goods, metals, &c.

Warehousemen and Shippers.	Ports.	Manufactures.
M. Darton, F., & Co., 72, St. John Street	San Francisco, Ceylon, and Mauritius	
Daunt & Senior, 27, Clement's Lane	India	Metals
Davidson, R. E., & Co., 6, Crosby Square	Spain, Canaries, and various	Linens, iron, machinery, earthenware, and general merchandise
Davidsons & Co., 9, Gracechurch Street	West Indies .	Hardware, machinery and general
Davis, D., & Sons, 15, Leadenhall Street	Various . .	Ferndale and 'Davis Merthys' steam coals
Davis, W. B., 106, Leadenhall Street	Bombay, Calcutta, Madras, China, and West Indies	Hardware, glass, cottons, Manchester goods, and general
De Lizardi, F., & Co., 124, Cannon Street	West Indies, America, and Spain	Linens, cottons, iron, medicines, glass, beer, and general
De Rass & Son, 17, Fenchurch Street	Cape of Good Hope, Australia, West Indies, &c.	Woollens, cottons, haberdashery, hardware, earthenware, wines, spirits, &c.
De Salamanca, J., 11, Leadenhall Street	Spain and various .	Iron, coals, railway plant, and general merchandise
De Salles, J., & Co., 28, Fenchurch Street	Brazil and River Plate	Cottons, linens, hardware, earthenware, glass, beer, ales, and machinery
Deacon, E. & A., 34, Rood Lane	Hong Kong, China, and India	Pig lead
Deane & Co., 46, King William Street	Australia and various	Ironmongery, &c.
Deane, J. & A., 2 and 3, Arthur Street East	East Indies and Colonies	Saddlery, harness, machinery, &c.
Deane, J., & Sons, 30, King William Street	Various . . .	Arms, &c.
Deare, F. D., & Co., 19, Coleman Street	Algoa Bay and Cape of Good Hope	
M. W. Defries & Sons, 147, Houndsditch	Sydney, New Zealand, Indies, China, Egypt, and Turkey	Hardware, chandeliers, glass, earthenware, &c.
Dennis Bros. & Co., 24, Martin's Lane	Continent and various	Tin plates, metals, &c.
Dent, Palmer & Co.. 11, King's Arms Yard	China and India .	Manchester goods, metals, &c.
Dickenson, Rose & Co., 3, King William St.	East India and China	Manchester goods, woollens, metals, and soft goods
M. Dixon, James, & Sons, 37, Ludgate Hill	India and various .	Electro-plate, Britannia metal, and shooting tackle
Dobree, G., & Sons, 6, Tokenhouse Yard	West Indies and various	Estate stores and machinery
M. Dolland & Co., 1, Ludgate Hill	Various . . .	Telescopes and optical instruments

Warehousemen and Shippers.	Ports.	Manufactures.
M. Donald, Altreg & Co., 33, Cornhill	India and various	Machinery and railway plant
M. Doukin & Co., Blue Anchor Road, s.e.	Colonies and various	Machinery
M. Dougall, J. D., 59, St. James Street, s.w.	India and Colonies	Guns, &c.
M. Doulton, Henry, & Co., 63, High Street, Lambeth, s.	Various	Pottery, bricks, and tiles
M. Doulton, J., Bros., & Co.,28, High Street, Lambeth, s.	Various	Earthenware
Duncan, George, & Co., 2, East India Avenue	East Indies	Manchester goods, metals, and general. (Ship insurance brokers)
Dunell, Henry J.	Cape	Hardware, soft goods, and general
Dunham, William, 48. Mark Lane	Australia, Cape of Good Hope, Canada, India, France, Germany, and Russia	Flour-mill machinery, mill-stones, hardware, and lubricators
Dunlop, George, & Co., 9a, New Broad Street	Continent, East Indies, and Ceylon	Woollens, machinery, hardware, colonial produce, and general merchandise
Dunn, W., & Co., 6, Lime Street Square	Cape, Natal, India, Algoa Bay, China, Australia, &c.	Manchester goods, woollens, hardware, and general
Durrant, A., 30, Great St. Helen's	West Coast of Africa, Callao, and India	Spirits, beads, earthenware, iron, grocery, and cotton goods
Duthie, W., 1, Wallbrook Buildings	Cape	Cottons, earthenware, and woollen manufactured goods
Dymes, Daniel D., 9, Mincing Lane	Madras	Hardware, machinery, iron, Manchester goods, &c.
Eames & Co., St. Michael's House, Cornhill	America	Iron
Earnshaw, Worsley, & Co., 17, Fenchurch Street	West Indies	Manchester goods, hardware, &c.
East 'India Railway Company, Moorgate Street	East Indies	Railway plant, engines, &c.
M. Ely Brothers, 254, Gray's Inn Road	East Indies, China, Japan, and various	Ammunition and arms
Elin, John, & Co., 3 and 4, Great Winchester Street Buildings	West Indies and United States	Cotton goods, provisions, arms, apparel, ammunition, and general
Elt, Charles, 4, St. Helen's Place	East Indies	Manchester and piece goods, metals, paints, oils, &c.

Warehousemen and Shippers.	Ports.	Manufactures.
Enthoven, H. J., & Sons, 17, Gracechurch Street, and Lead Works, Upper Ordnance Place, Rotherhithe	Mediterranean & South of Europe	Cottons, linens, colonial produce, and metals
Eschmann Brothers & Walsh, 12, St. Bartholomew's Square, St. Luke's	United States and Continent	India-rubber surgical instruments, &c.
M. Evans & Wormull, 6, Dowgate Hill	India, Australia, United States, &c.	Surgical instruments and appliances
W. Evered, R., & Son, 50, Bishopsgate Street Without	Australia, &c. . .	Hardware
M. Everitt, Allen & Sons, 118, Cannon Street	America, Russia, &c. .	Tube wire and metals
Fanfax, William Henry, 74, Little Britain	Various . . .	Hardware
Faithful, Cookson & Co., 9D, New Broad Street	Various . . .	Iron and iron ore
Falkland Islands Company, 39A, Gracechurch Street	Stanley, Falkland Isles	General merchandise, coals, &c.
W. Farmiloe & Sons, 34, St. John Street	Various . . .	Lead, glass, oils, and colours
W. Farrow & Jackson, 16, 17, and 18, Great Tower Street; 1, Harp Lane; 91, Mansell St.; and 8, Haymarket	Various	Iron wine bins, beer engines, bottling and corking machines, and all requisites for bottling wines and spirits
Faring & Co., 36, Queen Street	India and others .	Tin and Japan ware goods
Fenn & Ellis, 32, Botolph Lane	West Coast of South Africa, &c.	Cotton goods, provisions, ammunition, and general
Field, O., & Co., 76, Mark Lane	United States and North America	Drugs, dye stuffs, chemicals, colours, metals, paint, gums, East India produce, oil, oil-seed, and sundries
Findlay, Denham & Brodie, 31, Great St. Helen's	Australia, New Zealand, Cape Town, Algoa Bay, and Vancouver's Island	Iron, hardware, soft and piece goods, groceries, and general merchandise
Findlay, John. & Co., 10, Laurence Pountney Lane, Cannon Street	Beaunos Ayres	Machinery, hardware, printing material, and general merchandise
Fleming, Seymour & Co., 18, St. Helen's Place	Russia and Alexandria	Metals and general
Forbes, D. W., & Co., 40, Upper East Smithfield	Various . . .	Ship's ironmongery
Forbes, J. A., 8, Lime St.	East Indies and China	Hardware-earthenware
M. Fox, Walter, & Co., 13, High Holborn	Australia . . .	Wire-work sieves

Warehousemen and Shippers.	Ports.	Manufactures.
Fraser, D., 8, New Broad Street	Russia . . .	Iron, machinery, coals, &c.
Preeth, S., & Co., 60, Gracechurch Street	India, &c. . .	Manufactured iron
Froom & Co., 8, New Broad Street	Russia . .	Machinery and iron
Frost, J. E., 85, Goswell Street	Colonies, &c. .	Steam cocks and water gauges
Fry, James, & Co., Gresham House	Various .	Metals
Galbraith, H. J., & Co., 9, Crosby Square	China	Manchester goods, iron, glass ware, &c.
M. Gann, Jones & Co., 171, Fenchurch Street	Colonies .	Sewing machines, shirts, collars, clothing, &c.
Gaminara, & Co., 3, New London Street	South America, South of Europe, &c.	Manchester and Sheffield goods and general
Garden & Son, 200, Piccadilly	East Indies .	Arms, ammunition, saddlery, &c.
Gardner, Joseph, & Sons, 5, New London Street	South Russia and Turkey	Machinery and general
Gardner, R., 46, Lime Street	West Indies . .	Machinery and estate stores
Garnham, J.B., 10, Laurence Pountney Lane	Various . .	General merchandise, hardware, and machinery
Gavin, Alexander, 9, Moorgate Street	Baltic	Colonial produce, oils, and metals
Geiselbrecht, J. C., 8, Leadenhall Street	Baltic, North of Europe, Australia, New Zealand, Cape, East Indies, and China	Drugs, chemicals, metals, and general merchandise
Gellaty, Hankey, Sewell, & Co., 109, Leadenhall Street; and Manchester	India, China, and general	Cottons, woollens, hardware, and general
Gerber, Chrestien & Co., 59, Mark Lane	Rangoon, Akyal, &c. .	Manchester staple, coloured yarns, and corrugated iron
M. Gerich, Samuel, & Co., 22 to 30, Buttesland Street, N.	Colonies and various .	Patent spring hinges
Geelgud, H., 65, Gracechurch Street	Continent, &c. . .	Machinery, railway plant, and iron
Gillesby & Scott, 2, Brabant Court, Philpot Lane	West Coast of Africa, &c.	Coals
M. Gillett & Bland, Steam Clock Factory, White Horse Road, Croydon	Australia, China, India, South America, Turkey, Spain, Canada, Greece, Russia, &c.	Church, turret, stable, house, and musical clocks of every description; bells and patent chiming machines
M. Gillott, J., & Sons, 37, Gracechurch St.	Various . . .	Steel pens

Warehousemen and Shippers.	Ports.	Manufactures.
Ginham, John, J., 61, Moorgate Street	Various . . .	Iron and wood houses, wire netting, fencing, and hurdles
Gospel Oak Iron Company, 74, King William Street	Australia and various .	Galvanized iron
Gossell, Otto, 22, Moorgate Street	Continent . . .	Rails, iron, steel, and general
W. Graetzer & Hermann, 73, Aldermanbury	All parts of Europe, America and Colonies, East Indies, and China	Glass, lead, colonial produce, cottons, woollens, and linens
Grant, James (late Smith & Grant), 17, Gracechurch Street	East Indies, &c. .	Manchester goods, metals, hardware, and general
Grantoff, B. A., & Co.. Jeffry Square	North of Europe, Russia, South America, East Indies, and China	Machinery, hardware, and English manufactures in general
Gray, Beavis, & Co., 58, Lombard Street	Continent . . .	Coal-tar products, chemicals, and colonial produce
Gray, C. W., & W., 31, Great St. Helen's	West Indies and various	Cottons, linens, wearing-apparel, machinery, and estate stores
Green & Holland, 15, 16 & 17, Coal Exchange	Various . . .	Coals
Green, J., 35 & 36, Upper Thames Street	Cape and various . .	Earthenware, glass, and china
Green, Thos., 20, Great Winchester Street	Various . . .	Iron and hardware
Gregson & Co. (East India and China agents), 14, Austinfriars	East Indies, China, and others	Metals, cottons, woollens, linens, hardware, and general
Greig, G., 3, George Yard, Lombard Street	Cape of Good Hope and West Coast of Africa	Hardware, apparel, machinery, cottons, and general
Grice, Wm., & Co., 21, East India Chambers	Melbourne, Adelaide, New Zealand, Canterbury, Otago, and others	Soft goods, silks, beer, hardware, machinery, wines, spirits, boots and shoes, and general
Grieves, James, & Co., 43, Mincing Lane	Calcutta . . .	Metals in general
Griffith & Co., 84 and 133, Cannon Street	Iron to all parts of the world	All kinds of malleable iron, Scotch pigs, steam engines, machinery, and metals
Griffiths, Williams & Co., 2, Crown Court, Philpot Lane	Various . .	Copper and yellow metal
Grindlay & Co., 55, Parliament Street	East and West Indies and China	Hardware, machinery, cottons, and linens
Guest, Sir John, & Co. (Bart.), 13, King's Arms Yard	Various . . .	Iron

Warehousemen and Shippers.	Ports.	Manufactures.
M. Gwyne & Co., Essex Street, Strand	East and West Indies, Japan, China, Demerara, South America, and various	Machinery for works of irrigation, drainage, &c., for dry docks, canals, plantations, sheep washing, and manufacturing purposes
Hadden, James A., & Co., 25, Fenchurch Street	Ceylon and East Indies	Ironware and general goods
Hagedom, Frederick Wm., 150, Leadenhall Street	Hong Kong and China	Manchester and woollen goods, iron, and general
Hall & Holtz, 6, St. Benet's Place	China	Apparel, machinery, cottons, &c.
Hall, Edward James, 144, Minories; and 6, America Square	West Indies, Vincent, and various	Ironmongery and machinery
Hammond, C., 6, Billiter Square	West Indies . .	Machinery, guano, estate stores, and general
M. Hampson & Bettridge, 47, 48 & 49, Old Bailey	All ports . . .	Bookbinder's machines, presses, and tools
M. Handyside, A., & Co., 32, Walbrook	East Indies, China, Japan, Russia, Sweden, Norway, and the Colonies	Iron bridges, roofs, engines, boilers, and all classes of iron work
Hanson, Henry Allix, 23, Great Winchester Street	Constantinople . .	Machinery and general
Harbottle, J., 1, Alderman's Walk	Cuba, Spain, Russia, and Mediterranean	Iron, coals, colonial produce, and general
Hardford & Bristol Brass Battery and Wire Company, Dowgate	Various .	Ingot and sheet brass, copper, and brass wire
Hanaden, S., & Co., 3, Chapel Place, Poultry	India	Stationery, books, plated ware, and bronze goods, and general merchandise
M. Harrild & Sons, 20, Farringdon Street	Various .	Printing presses, machines, type, and printing materials
Harris, Scarfe & Co., 28, Martin's Lane, Cannon Street	Adelaide, &c. .	Manchester and Sheffield goods and general
Harrison, J., & Co., Skinner Place, Size Lane	Various . .	Manchester and Bradford goods, iron, and machinery
Harrold Brothers, 32, Great St. Helen's	Adelaide . . .	Hardware, &c.
Harvey, Brand, & Co., 37, New Broad Street	East Indies, China, and Japan	Cottons, woollens, linens, metals, &c.
Harward, W. T., & Co., 2, East India Avenue	America . . .	Scrap-iron. (Shipbroker)

Warehousemen and Shippers.	Ports.	Manufactures.
Harwin, R. (at Harvey & Greenacre's), 85, Gracechurch Street	Cape, Natal, and West Indies	Woollens, cottons, haberdashery, and hardware
M. Hasluck, L., George Yard, Lombard Street	Colonies and India	Watches and clocks
Herdeman, Hayton & Co., 13, Old Jewry Chambers	Cape	Fancy goods, general merchandise, machinery, and drapery
Heuitzmann & Rochusen, 23, Abchurch Lane	Africa and West Coast	Cotton goods, apparel, hardware, and provisions
Held, Charles, & Co., 63, Great Tower Street	America and Continent	Metals, cottons, &c.
Henderson, George, 7, Mincing Lane	East Indies, China, and others	Metals and Manchester goods
Henderson, George, 58, Lombard Street	Leghorn, &c	Iron, coal, paints, and various. (Ship insurance broker)
Hendewerk, R., 79, Mark Lane	France, Baltic, and North of Europe	Glassware, iron, cement, and provisions
Heny, Michael, 7, South Place, Finsbury	Jamaica, West Indies, and South America	Hardware, cottons, linens, &c.
M. Hernulewicz & Co., 43, Fish Street Hill	Colonies	Iron manufactures for farming purposes, and fencing wire
M. Heulett & Co., 55 and 56, High Holborn	Rio, Singapore, &c.	Gas fittings
Hezerdahl, Schonberg & Co., 10, Cornhill	America	Iron and metals
Hickie, Borman & Co., 127, Leadenhall Street	Continent, East Indies, China, &c.	Coals, iron, soft goods, and general
Higgin, E., & Co., 2, East India Avenue	East Indies and China	Ironwork, woollens, apparel, cottons, beer, machinery, hardware, and railway materials
M. Higgs & George, 60, Cannon Street	Various	Ironmongery
M. Hill & Smith, 97, Cannon Street, and Staffordshire	Australia, New Zealand, and Cape	Iron and wire fencing, hurdles, and gates of every description
Hill, Richardson & Wright, 35, Great St. Helen's	West Indies, Australia, Spain, and various	Metals, manufactured goods, and general
Hillier & Co., 46, Lime Street	Continent, United States and Australia	Hardware, firearms, cotton goods, and general
Hinton Bros. & Co., 80, Old Broad Street	Mediterranean and America	Iron and various
Hitchcock & Co., 38, St. Mary Axe	Continent	Metal
Horne & Thornthwaite, 122, Newgate Street	North and South America	Philosophical instruments
Houghton, Smith & Co., 2, Jeffrey's Square	East Indies and various	Hardware and Manchester, &c.
Howard, Alfred, 3, Leadenhall Street	Various	Yellow metal, zinc, iron, tin plates, and copper

Warehousemen and Shippers.	Ports.	Manufactures.
Howden, Alexander & Co., 19, Birchin Lane	East Indies and South America	Machinery and metals
Howe Machine Co., 64, Regent Street	East Indies, Australia, and various	Sewing machines
Hubband, John, & Co., 4, St. Helen's Place	Russia	Machinery, iron, and general
Hughes Brothers, 30, Gracechurch Street	Continent	Hardware and general
Hughes, Henry, 83, Gracechurch Street	Various	Metals
Hughes, J. W., 24, Leadenhall Street	United States and various	Metals and tin plate
Holworthy, J. M., & Co., 30, Great St. Helen's	Australia and various	Hardware
Hooper, A. D., & Co., 40, Old Broad Street	China and Japan	Metals and hardware
Hooper, John R., & Co., 32, Great St. Helen's	East Indies, China, and Japan	Manchester and Bradford goods, metals
Hoperaft & Broadwater, 3, Billiter Square	Australia, New Zealand, China, &c.	Manchester goods, iron, and general
Hurndall, H. L., 2, Skinner's Place, Size Lane, Bucklersbury	New Zealand	Agricultural implements, Manchester goods, hardware
Ibbotson Brothers, 1, Skinner's Place, Size Lane, Bucklersbury	Various	Steel and iron files, and engineers' and other tools
Ilbery, J., 23, Beer Lane	Australia, America, Cape, India, China, Japan, Canada, and various	Oilman's stores, Manchester goods, hardware, and general
Irving, T., & Co., 17, Gracechurch Street	Australia, Monte Video, New Zealand, and various	Manchester goods, hardware, metals, and general
James & Shakspeare, 10, Austin Friars	Australia, New Zealand, East Indies, China, and United States	Metals
James, Walter, 10, St. Benet's Place, Gracechurch Street	East and West Cape, China, Australia, and various	Machinery and general
Jamieson, Wm., & Co., 9, Fenchurch Street	East Indies, China, and Japan	Manchester goods, metals, machinery
Johnson Brothers, & Co., 6, Waterloo Place, Pall Mall	Various	Iron fences, &c.
Johnston, Charles, & Co., 150, Leadenhall Street	Brazils and North America	Machinery, metals, and general
Johnson, Matthey, & Co., 77, Hatton Garden	English Colonies	Metals, &c.
Jones, Scott & Co., 62, Basinghall Street	Melbourne	Hardware and general
Joshua Brothers & Co., 9, Great Winchester Street	Melbourne and various	Metals, &c.

Warehousemen and Shippers.	Ports.	Manufactures.
Jourdain & Co., 10, Austin Friars	East Indies, Mediterranean, South of Europe, and various	Pig iron and general merchandise
Kaltenbach & Schuntz, Alderman's Walk, and Liverpool, Hamburg, and Bordeaux	East Indies, China, and Japan	Machinery, metals, and general
Kerr, George, & Co., 6, Great Winchester Street Buildings	Australia and New Zealand	General
Kirk, James, & Co., 27, Mincing Lane	Spain .	Railway materials and general merchandise
Kleeberg, Martin, 21, Throgmorton Street	Continent .	Bullion
Knowles & Foster, 42, Moorgate Street	Portugal, Brazils, and China	Arms, &c.
Krauss, Klein Paul, 15, Tower Hill	Continent and various .	Guns and engineer's stores
Kreeft, Howard & Co., 124, Fenchurch Street	Continent .	Hardware and machinery
Kruger, Aug., 34, Throgmorton Street	Valparaiso .	Iron ware, &c.
Kuhuer, Hendschel, & Co.,145,CannonStreet	West Indies.	Machinery, &c.
Laird, J. W., 6, Bishopsgate Street Without	Wellington, Australia, India, and China	Arms, &c.
Lampson, C. M., & Co., 64, Queen Street	United States .	Railway metals
Lang, J., & Co., 32, New Broad Street	Cape .	Arms, &c.
Lanyon, J. C., 48, Gresham House	Australia .	Machinery and hardware
Lawrence, Clark & Co., Windsor Chambers, 20, Great St. Helen's	East Indies, China, Australia, and Honolulu	Hardware, British manufactured goods, and general merchandise
Lawson, P., & Son, 2, Budge Row, Cannon Street	Australia, New Zealand, West Indies, Cape Town, France, Germany, &c.	Agricultural and horticultural implements, &c.
Lazard, E., & Co., 20, King's Arms Yard, Moorgate Street	Baltic .	Machinery, hardware, &c.
Lazarus, Lewis, & Sons, 29, Great St. Helen's	India, &c. .	Metals
Leaver & Breege, 25 & 26, Houndsditch	Colonies and various .	Birmingham and Sheffield goods
Leaver, J. R., 16, Water Lane	West Coast of Africa .	Manchester, Sheffield, and general goods
Leddell, Henry, & Co., 24, High Holborn	Indies, Australia, New Zealand, Cape, and Continent	Britannia metal, &c.
Leech, Harrison & Forwood, 30, Great St. Helen's ; Liverpool and Manchester	East Indies, China, Japan, Australia, &c.	Manchester goods, hardware, &c.

Warehousemen and Shippers.	Ports.	Manufactures.
Lent. J. O., 106, Cannon Street	Various · · ·	Manufactured iron, &c.
Lepage, R. C., & Co., 1, White Friars Street	Calcutta · · ·	Machinery, &c.
Leon, Lewis, 23, New Broad Street	New York and West Indies	Hardware and general
Levich, Joseph, & Son. 9, King's Arms Yard	Cape and Algoa Bay	Hardware
Levich, James, 9, King's Arms Yard	Sydney ·	Hardware
Levin, M. L., 1, Bevis Marks	Africa · · ·	Fire-arms, &c.
Lias, H. J., & Sons, 7, Salisbury Court, Fleet Street	Australia, East and West Indies, Cape, &c.	Plated ware and cutlery, &c.
Lindo, Daniel, 35, Great St. Helen's	West Indies · ·	Iron, hardware, &c.
Lindley, Arthur, & Co., 31, Great St. Helen's	Spain · · ·	Hardware and general
Linnington, A. H., 58, Fenchurch Street	Nevis and St. Kitts ·	Iron and general
Locke, Lancaster & Co., Peter's Chambers, Cornhill	America, Calcutta, East and West Indies, and China	Tea, pig, white, red, and sheet lead, and pipes
Lockhart, Toger & Co., 3, Storey's Gate, Westminster	Continent and various ·	Wrought iron tubes, chains, anchors, and heavy iron goods
London Warming and Ventilating Company (Lim.), 23, Abingdon Street, Westminster	Trieste, Paris, Brussels, St. Petersburgh, &c.	Stoves, furnaces, and warming apparatus
Lonergan F. W., 22, Lime Street	Central America · ·	Iron, &c.
Low Moor Iron Company, 98, Cannon Street	Copenhagen, Amsterdam, and Hong Kong	Iron, bars, and plates
Macfarian & Co., 17, Langbourn Chambers, Fenchurch Street	East India and China · · ·	Machinery, hardware, and general
Macfarlane, W., & Co., 84, Upper Thames Street	India and Australia · ·	Ornamental cast-iron work, rain-water castings, &c. ·
Mackay, Gellier & Co., 1, Leadenhall Street	Australia · · ·	Metals and general
Mackaught, Robertson & Co., 1, Bankend, Bankside	Various · · ·	Iron sheets and plates
Madox & Co., 36, Mark Lane	Mediterranean and South America	Metal ware, machinery, &c.
Malcolm, Brunker & Co., 14, St. Mary Axe	India, China, and Japan	Manchester goods, metals, and general
Manning & Co., High Holborn	Australia · · ·	Iron fencing, hurdles, and general merchandise

Warehousemen and Shippers.	Ports.	Manufactures.
Mappin & Webb, 71 & 72, Cornhill	India, 'Australia, and various	Cutlery, &c.
Marshall & Co., Sussex Place, Leadenhall Street	New Zealand and Africa	Ironmongery
Matveieff. C., & Co., 32, Great St. Helen's	China and Russia	Machinery, &c.
Mavro, Valieri & Co, 112, Gresham House	Russia, Danube, Levant, and South of Europe	Metals, hardware, &c.
Malwell, Robert, & Co., 9, Mincing Lane	Madras Coast	Iron and Manchester goods
Maxondoff, A., 14, Union Court, Old Broad Street	Russia, Mediterranean, and United States	Metals, &c.
May, G., & Co., 17, Finsbury Circus and 1, East India Avenue	East Indies .	Manchester goods, metals, &c.
McAndrew, W., & Sons, 85, Gracechurch Street	Spain, Azores, Portugal, and Vauture Ports	Hardware, iron goods, agricultural implements, &c.
McArthur & Co., 23, Rood Lane	Various	Metals and iron
McComas, Thomas, & Co., 55, Old Broad Street	Port Phillip	Hardware, metals, &c.
McEvan, James, & Co., 122, Cannon Street	Melbourne, Sydney, and New Zealand, part	Iron, hardware, &c.
McMaster, J., 25, East-cheap	Cape .	Hardware, &c.
Meadows, T., & Co., Milk Street	United States, Australia, and other ports	General merchandise, railway plant, machinery, &c.
Mears and Stainbank, 267, Whitechapel Road	Canada, United States, America, West Indies, Australia, India, and all parts	Bells
Meier, C. G., & Co., 3, Brabant Court, Philpot Lane	China, East Indies, and various	Machinery, hardware, &c.
Menasce, T. L., Sons, & Co., 1 and 2, Great Winchester Street Buildings	Egypt .	Manchester, Birmingham, and general goods
Merryweather & Sons, 63, Longacre, and York Street, Lambeth	Various	Fire engines and general machinery
Merton, H. R., & Co., 117 and 118, Leadenhall Street	East Indies, Australia, and various	Metals, hardware, &c.
Michell, R. R., & Co., 2, Crown Court, Philpot Lane	Various	Block, bar, and other tin
Miles Brothers & Co., 79, Gracechurch Street	Australia, New Zealand, and Colonies generally	Hardware, &c.

Warehousemen and Shippers.	Ports.	Manufactures.
Miles, Gould, Druce & Co., 29, Upper Thames Street	Various	Iron
Millar, W., 49, Fenchurch Street	Mediterranean and Australia	Hardware, machinery, iron, lead, and shot
Montefiore, John, 4, St. Benet's Place, Gracechurch Street	West Indies and America	Hardware, machinery, &c.
Montigny, Reini De, 106, Cannon Street	China	General
Moore & Manby, 3, Billiter Square, and Dudley	China, Australia, and various foreign governments	Iron
Morton, J., & Co., 6 and 7, Sherborne Lane	China, New Zealand, Australia, Japan, and various	General, hardware, iron, and metal
Morewood & Co., Leadenhall Street	Various	Galvanised iron roofs, sheet iron, &c.
Morgan, T., & Son, New Street, Bishopgate Street	Sierra Leone, West Coast of Africa, &c.	Native implements, arms, &c.
Morrison, A., & Co., 10, Austin Friars	New Zealand and West Indies	British manufactured goods, metals, &c.
Mort, W., & Co., 155, Fenchurch Street	Sydney	Machinery, iron, &c.
Mortimer, H. H., & Co., 10 and 25, Bush Lane, and Birmingham	Hong Kong, Australia, &c.	Hardware
Morum Brothers, 1, Guildhall Chambers	Cape	General
Muntz Metal Company, 23, Rood Lane	Various	Metals
Murdoch, H. H., 30, Great St. Helen's	East Indies	Machinery, hardware, &c.
Naoroji, Dadabhai & Company, Great St. Helen's	India	Manchester goods, machinery, &c.
Nash, Samuel, & Co., 137, Fenchurch Street and at Cardiff	United States	Metals
Naylor, Benyon & Co., Palmerston Buildings, Old Broad Street	Continent and United States	Railway materials, iron, steel, &c.
Neilson, C., & Sons, 12, Great St. Helen's	West Indies and various	Iron, &c.
Netter, Charles, & Co., 31, Bush Lane	Egypt and Syria	Machinery, Birmingham goods, iron, tools, &c.
Newby, E. H., 39A, King William Street	East Indies	Machinery and rifles
Nicholls, T., & Co., New City Chambers, Bishopsgate Street	Cape and various	Hardware, &c.

Warehousemen and Shippers.	Ports.	Manufactures.
Nicolson Brothers, 5, Jeffrey's Square	Australia, New Zealand, Rio De Janeiro, Calcutta, Madras, Bombay, Colombo, Yokohama, Hong Kong, Cape Town, and Natal	Hardware
Norrington, Pitts & Co., 23, Great St. Helen's	Various . . .	Iron
Novelli & Co., 2, Crosby Square	Spain, Egypt, and United States	Metals, &c.
Oakes Brothers & Co., 26, Nicholas Lane	Madras, East Indies, China, &c.	Hardware, &c.
Oakes, T., & Co., 10, Austin Friars	East Indies and China	Hardware, metals, &c.
Orme, Frederick, Ethelburgha House, 70 and 71, Bishopsgate Street Within	Continent . . .	Hardware and machinery
Osborne, C. S., Great Garden Street, Whitechapel	New York and Moscow	White metal and brass castings
Owen, Richardson & Co., 3, Newman's Court, Cornhill	Baltic and North of Europe	Machinery, hardware, &c.
Palmer, Edward S., 8, Cullum Street	East Indies, China, South America, &c.	Machinery, arms, &c.
Palmer, J. N., 22, Leadenhall Street	Australia . .	Ironwork and general
Panulcillo Copper Company (Limited), 25, Great St. Helen's	Coquimbo . .	Tools and machinery
Parkes, J., & Co. (Lim.), 152, Upper Thames Street	Various . .	Hardware
Parry, Lovel & Co., 122, Cannon Street	Continent and Africa .	Iron, tin-plate, and general
Pavia, Charles, 83, Lower Thames Street	Italy . .	Tin-foil
Peacock, C. G., 1, Bishopsgate Street	North and South America	Cutlery, &c.
Peal & Chattoch, 149, Upper Thames Street	Various .	Metal, tin-plates, &c.
Perkins, E., & Co., Lombard House, George Yard, Lombard Street	America .	Metals, &c.
Peters, G. D., & Co., 9, Bunhill Road	Australia, West Coast of Africa, West Indies, China, and Japan	Hardware, machinery, and engineers' stores
Picard, Joseph, & Co., Coffee Planter's Hall, Farringdon Street	Colonies and various .	Coffee machines

Warehousemen and Shippers.	Ports.	Manufactures.
Pini, J., Roncorini Brothers, 22, College Hill	River Plate and South America	Hardware, &c.
Pinto, Leete & Nephews, 24, Moorgate Street	Spanish and Portuguese Settlements, Brazils, and West Coast of Africa	Machinery, hardware, &c.
Pothohier, Tilsley & Co., 150, Leadenhall Street	Cape, Brazils, Natal, and various	Machinery, railway plant, and general
Powis, Charles, & Co., 69, Gracechurch Street	Australia, East and West Indies, South America, and various	Sewing and general machinery, engines and boilers
Preeston, H. A., & Co., 34, Fenchurch Street	Portugal, Spain, and Mediterranean	Iron, &c.
Previte & Greig, 3, Newman's Court, Cornhill	Spain	Cornhill metals, machinery
Price Brothers, 15, New Broad Street	East Indies, China, Australia, New Zealand, and South America	Metals, &c.
Pulling, R. W., 11, St. Bennil Place	China and various .	Iron and tin plates
Quincey, Harcourt, 5, Bond Court, Walbrook	Australia, Cape, East and West Indies, Russia, Spain, South America, and West Coast of Africa	Ironmongery, cutlery, hardware, Britannia metals, cast and wrought iron tubes, &c.
Radeke, Arthur C., 15, Fish Street Hill	Italy	Hardware and general merchandise
Raphael, R., & Sons, 25, Throgmorton Street	Various . . .	Bullion
Ravene, Jacob, & Sons & Co., 45 and 46, Great St. Helen's	Various . . .	Metals
Redruth Tin Smelting Company, Golden Heart Wharf, Dowgate	Various . . .	Ingot and sheet brass, copper and brass wire
Rehder & Co., 2, Lime Street Squar	China and Brazil. .	Metals, hardware, and general
Relley, E. M., & Co., 502, New Oxford Street	Australia, New Zealand, South America, West and East Indies, Japan, and China	Arms, &c.
Retter, Hugh, & Co., 14, South Street, Finsbury	Australia and various .	Hardware
Reuss, Ernest, & Co., 39, Lombard Street	India, Russia, South America, and Continent	Machinery and railway plants of every description
Richardson, Brothers & Co., 12, St. Helen's Place	Australia and New Zealand	Metals, agricultural implements, hardware, machinery, and general

Warehousemen and Shippers.	Ports.	Manufactures.
Richardson, E., & Co., Sharp's Wharf, Wapping	India, China, Australia, and America	Hardware, hollowware, and every description of iron and steel goods
Rixon, Alfred, & Co., 35, Eastcheap	Cape and Java .	Hardware and Manchester goods
Robertson, J. R., 2, Billiter Square	East Indies . .	Manchester goods, &c.
Robinson, Fleming & Co., 21, Austin Friars	Russia Baltic and North of Europe	General merchandise
Rock, T. Dennis & Co., 46, Leadenhall Street	East Indies, China, Japan, Australia, and various	Metals, hardware, and general merchandise
Rodocanachi, Sons & Co., 37, Threadneedle Street	Various . . .	Soft goods, machinery, and general sundries
Rogers, John, & Co., 134, Leadenhall Street	Australia, New Zealand, America, India, China, and various	Hardware, machinery, metals, and general goods
Rosselli, A. & E., 32, Fenchurch Street	Trieste, Venice, Mediterranean, and South of Europe	Copper, &c.
Rouquette & Co., 1, Crosby Square	Mexico, United States, Continent, Spain, and various	Hardware, machinery, and general
Rownson, Drew & Co., 217, Upper Thames Street	Various . . .	Iron, steel, zinc, nails, and hardware of all descriptions
Runciman & Scott, 5, Laurence Pountney Lane	United States and various	Galvanized iron, &c.
Russell, John, & Co., 69, Upper Thames Street	Various	Boiler and gas tubes
Russian Trading Company, 39a, Gracechurch Street	Russia .	Iron, machinery, and general
Sanderson & Co., 14, St. Helen's Place	East India, China, and various	Metals, Manchester goods, and general merchandise
Sanford & Bird, 4, Cullum Street	Australia and America	Iron, tin plates, and metals
Sassoon, Davis & Co., 15, Leadenhall Street	East India, China, and Japan	Manchester and Bradford goods, metals, and general
Savage & Co., 42 and 43, Eastcheap	Spain, Australia, an Belgium	Machinery
Schiller, F., 4, St. Helen's Place	Calcutta . .	Hardware and machinery
Schesinger, W. J., & Co., 16, Finsbury Street	Continent . .	Hardware
Schroder & Co., 5, Abchurch Yard	Various . . .	Iron and iron manufactures
Schuster, Son & Co., 90, Cannon Street	All parts . .	General and metals

Warehousemen and Shippers.	Ports.	Manufactures.
Searle, Edward, 79, Gracechurch Street	Russia. . . .	Hardware and general
Selig, M., jun., 70 and 71, Bishopsgate Street Within	Continent . . .	Hardware and machinery
Shand, Mason & Co., 75, Upper Ground Street, Blackfriars	India, China, and Colonies	Fire engines, fire escapes, with their appliances, &c.
Shanks, A., & Son, 27, Leadenhall Street	Australia, East Indies, and South America	Lawn mowers, hoisting machines, &c.
Shaw & Thompson, 150, Leadenhall Street	United States, Russia, and Egypt, &c.	Iron and iron manufactures, &c.
Shaw, Charles, & Co., 4, Copthall Street	China, &c. . . .	Manchester goods, metals, &c.
Shaw, J., & Sons, 33, Cannon Street	Various . . .	Hardware
Shaw, Savill & Co., 34, Leadenhall Street	New Zealand . .	Machinery, iron, lead, &c.
Sheppard, J. & R., 106, Leadenhall Street	Continent, India, and Colonies	Hardware, machinery, &c.
Sheriff, Lindsay & Co., 3 and 4, Great Winchester Street Buildings	East Indies . . .	Hardware
Siemens Brothers, 3, Great George Street	India, Australia, Egypt, Cape Colonies, New Zealand, North and South America, and various	Telegraph cable posts (iron), &c.
Sierte, King, Drop & Co., 65, Fenchurch Street	Various . . .	Iron, &c.
Sims, Willyams & Co., 1, Queen's Street Place	Various .	Copper, yellow metal, and sheathing
Sinclair, Peters & Co., 2, East India Avenue	All parts . .	Railway material, iron work, machinery, and general merchandise
Sleeman, H. B., 106, Leadenhall Street	East India and others .	Manchester goods, metals, &c.
Smith, Edward A., & Co., 150, Leadenhall Street	China, Calcutta, East Indies, and Japan	Hardware, &c.
Smith, Fleming & Co., 18, Leadenhall Street	East Indies and China	Manchester goods, hardware, and general
Smith, Gilead A., & Co., Bartholomew House, Bartholomew Lane, Threadneedle Street	America . . .	Railroad iron, old rails, Bessemer rails, &c.
Smith, Jas., & Co., Dunster House, Mincing Lane	East Indies, China, Australia, and others	Manchester goods, machinery, and general
Smith, R., & Co., 4, New Broad Street	India and Continent .	Steel
Smithers, Fred. Oldershaw, 37, Lime Street	Buenos Ayres, Spain, and various others	Railway iron, &c.

Warehousemen and Shippers.	Ports.	Manufactures.
Smithall, W. P., 27, Mincing Lane	Palermo . . .	Hides, &c.
Snellgrove & Leech, 33, Mark Lane	Egypt, Russia, and Italy	Metals, colonial and British manufactures
Spaeth, Gus. L., & Co., 78, Mark Lane	Continent and America	Iron
Spartali & Co., 25, Old Broad Street	Mediterranean, South of Europe, West Coast of Africa, Russia, and Danube	Iron manufactures and Manchester goods
Speechly & Nicholson, 74, Little Britain	Colonies and South America	Sewing machines, hardware, and general merchandise
Spence, J. B., & Co., 75, Mark Lane	Various .	Metals
Spence, P. W., & Co., 17, Gracechurch Street	Australia and India .	Metals
Spencer, John, & Sons, 124, Fenchurch Street	Various . .	Hardware
Stephens & Reynolds, 31, Great St. Helen's	America . .	Railway materials
Stewart, Louis, Poultry	India . . .	Electro-plate jewellery
Stiebel, S., & Co., 19, Abchurch Lane	East Indies and West Indies, Mauritius, Cape, France, and Wallachia	Arms, hardware, machinery, and merchandise of every description
Stowe, H., & Co., 16, George Street, Mansion House	East Indies, China, Canada, and others	Metals, machinery, saddlery, hardware
Stovell & Brown, 9, Lime Street	Australia, East and West Indies, and various	Machinery, metal, estate stores, British manufactured goods, and general
Strange, Alderson & Co., 3, St. Helen's Place	East Indies and China	Metals, machinery, hardware
Strode & Co., 67, St. Paul's Churchyard	Various .	All kinds of gas-fittings
Sussman, J., & Co., 12, Size Lane	Continent and United States	All classes of manufactured and raw materials
Swanzy, F. & A., 122, Cannon Street	Cape Coast Castle, and West Coast of Africa	Hardware goods
Tallack, F., Birchin Lane	Colombo and East Indies	Metals, general machinery, Manchester goods, and general
Tangye Brothers & Holman, 10, Laurence Pountney Lane	Australia, New Zealand, East and West Indies, and various	Steam-engines, steam and other pumps, pulleys, blocks, lifting-jacks, and general machinery
Tanner, Richards & Sons, 23, Thornhill Place, Caledonian Rd.	Various . . .	Platina-pointed pens
Taylor, Jas., & Co., 82, Mark Lane	Various . .	Machinery

Warehousemen and Shippers.	Ports.	Manufactures.
Taylor, R. P., & Co., Adelaide Place, London Bridge	Colonies, &c. . .	Agricultural machinery
Temperley, John, 155, Upper Thames Street	Colonies, &c. . .	Hardware
Tennent, Charles, Sons & Co., 9, Mincing Lane	West Indies and various	Sugar machinery, &c.
The General South American Company (Lim.), 10, Palmerston Buildings	North and South America	Machinery and general
The Gospel Oak Galvanised Iron and Wire Company, 56, Upper Thames Street	Australia, New Zealand, and various	Galvanised iron wire
The Wanzer Sewing Machine Company (Limited), 4, Great Portland Street	Australia, Canada, and others	Sewing machines
Thomas, R., & Co., 40, Gracechurch Street	Various . . .	Hardware
Thompson, Arthur K., 31, Great St. Helen's	Mediterranean, South of Europe, West Indies, and South America	Machinery
Thomson, J., Borar & Co., 57½, Old Broad Street	Russia . . .	Machinery, &c.
Tibbats & Sons, 44 and 45, Bishopsgate Street Without	Madagascar, Australia, and various	Iron and brass bedsteads
Tiden, Nordenfelt & Co., 34, Clement's Lane	Sweden, Russia, Germany, United States, East Indies, &c.	Railway materials, metals, &c.
Till, E. D., 26, Lombard Street	Various . . .	Rails, bars, and tinplates
Trantmann & Co., 1, New Broad Street Court	China and Japan . .	Manchester and foreign manufactured goods, metals, and general
Treggon, Hickson & Co., 21, 22, and 23, Jewin Street	Colonies, various . .	Galvanised zinc and iron
Tucker, J., 3, Laurence Pountney Place, Cannon Street	Colonies and various	Machinery, hardware, and general merchandise
Tuckness, C., & Co., 1, Lime Street Square	South America and West Indies	Hardware, saddlery, &c.
Turton, T., & Son, 35, Queen Street	Various . . .	Sheffield goods, steel, and files
Twycross, J., & Co., Brentford, Middlesex	Australia . . .	Metals and general
Tyler, Hayward & Co., 84 and 85, Upper Whitecross Street	Various .	Machinery
Tyler, J., & Sons, 2, Newgate Street	India and various .	Hydraulic machinery, pumps, and engines

Warehousemen and Shippers.	Ports.	Manufactures.
Tyler, John Henry, 36 and 38, Abbey Street, Bermondsey	Australia, Cape Colony, New Zealand, Mauritius, East and West Indies	Machinery for raising-pumps, screws, hydraulic presses, &c.
Tyser, Y. D., & Co., 3, Crosby Square	East Indies, China, and Australia	Machinery, hardware, &c.
Ullmann, Herschhorn & Co., Fountain Court, Aldermanbury	Calcutta and Hamburg	Plated ware and general merchandise
Ullmer, Fred., 15, Old Bailey	Various . . .	Printing machines, presses, type, and printing materials
Van Drunen, Juan, 13, Crutched Friars	Continent, China, Japan, and others	Metals, &c.
Van Weede, Jacques G., 41, Gt. Tower Street Buildings	Spain, Portugal, Mexico, and various	Machinery, &c.
Vardon, John, & Co., 3, Gracechurch Street	Various . . .	Hardware
Vaughan, W., & Co., 19, St. Helen's Place	British Honduras, Gulf of Mexico, Nicaragua, and West Coast of Africa	Metals, hardware, arms, and general merchandise
Vicker, Sons & Co. (Lim.), 67, Palmerston Buildings	America, Russia, &c. .	Steels
Vivian & Sons, 3, Bond Court, Walbrook	China, India, Australia, Africa, and various	Yellow metal and copper
Vivian, Younger & Bond, 117, Basinghall Street	Sundry . . .	Metals
Von Dadelszen & North, 4, East India Avenue	Various . . .	Metals
Walker, Andrew G., 10, Laurence Pountney Lane	China and Japan . .	Metals and general merchandise
Wallace Brothers, 8, Austin Friars	East Indies . .	Metals, &c.
Ward, Richard, 1, New Broad Street	Italy	Machinery
Wardrop, Robert, 84, Lombard Street	Spain . . .	Hardware, &c.
Warner, John, & Sons, 8, Jewin Court	Continent and Colonies	Bells, &c.
Warner, Walduck & Co., 11, Old Jewry Chambers	Hong Kong and various	Metals
Warren, D. & T. G., 75, Old Broad Street	East Indies . .	Ironworks, hardware, machinery, &c.
Warrington Wire Iron Company (Lim.), 61, King William Street	Various . . .	Manufactured iron of all descriptions
Warrington Wire Rope Works (Lim.), 7, Great Winchester Street Buildings	Various . . .	Wire ropes

Warehousemen and Shippers.	Ports.	Manufactures.
Watt, Jas., & Co., 18, London Street	Various . . .	General machinery
Wattenbach, Hulger, & Co., 22, Great St. Helen's	East Indies and various	Manchester goods and metals
Waydelin, C., 148, Fenchurch Street	Continent and various .	Iron and general
Wedlake, M. & C., 118, Fenchurch Street	Colonies, &c .	Agricultural machinery
Wheeler & Wilson, Manufacturing Company, 139, Regent Street; and 43, St. Paul's Church Yard	All over the world .	Sewing machines
Whetham & Sons, 39A, Gracechurch Street	America and India .	Machinery, &c.
Whight & Mann, 143, Holborn Bars	Australia, New Zealand, East Indies, and China	Sewing machines
White, Child & Co., 22, College Hill	Continent . . .	Machinery and general merchandise
Whitfield, George, & Co., 17, Gracechurch Street	West Indies, South America, and various	Hardware, &c.
Whitmore & Bunyon, 28, Mark Lane	Various . . .	Machinery
Whitwell, Wm., & Co., 26, Lombard Street	Continent and various .	Pig iron, bars, rails, &c.
Whyte, Robert, 21, Duke Street, Aldgate	Cape '. . . .	Hardware and general
Wilkinson, Barkworth & Co., 16, Austin Friars	West Indies and various	Hardware and general merchandise
Wilcox & Gibbs, 135, Regent Street; and 150, Cheapside	Various .	Sewing machines
Williams, Foster & Co., 27, Clement's Lane	Various .	Copper and other metals
William, Harvey & Co., 216, Upper Thames Street	Various .	Block and bar tin
William Brothers, 42, Cannon Street	Africa, South America, and various	Rifle guns and pistols, revolving and breech-loading fire arms
Williamson, Geo., & Co., 2, East India Avenue	India	Hardware and tin pails
Wilson, Calder & Co., 30, Albert Buildings, Queen Victoria Street	Australia, East and West Indies, and various	Machinery, &c.
Wilson, Swale & Co., 147, Leadenhall Street	Mauritius .	Hardware and general
Wippermann, Gustave, 19, Water Lane	Various .	Metals

Read " Important Hints to Iron Masters and Merchants," at page [160],

which have reference to the whole List of Buyers and Shippers.

Warehousemen and Shippers.	Ports.	Manufactures.
Wood, W. A., 77, Upper Thames Street	Australia, New Zealand, Cape, Continent, and various	Mowing and reaping machines
Woodman, T. H., & Co., 34, High Street, Borough	Various .	Agricultural machinery
Wright, A., & Co., 16, Little Alie Street, Whitechapel	New Zealand, Australia, and Colonies	Smiths' and house bellows and portable iron forges
Wyche, C., 3 and 4, Skinner's Place, Size Lane	Holland and Java .	Manchester goods, hardware, saddlery, &c.
Wylie, Allan C., & Co., 84, Cannon Street	Brazils, West Indies, Australia, Japan, and Continent	Machinery and metals

LIST

OF ALL THE

BLAST FURNACES IN THE UNITED KINGDOM.

SHOWING ALL NOW IN BLAST AND THOSE STANDING IDLE, WITH
THE NAMES OF THE WORKS AND THEIR PROPRIETORS.

Name of Works and Owners.	Built.	In Blast.	Kinds and Quality of Iron made.
SOUTH STAFFORDSHIRE & EAST WORCESTERSHIRE.			
WOLVERHAMPTON.			
Chillington—Chillington Iron Co.	6	4	Good grey forge
Parkfield—Parkfield Iron Co.		$3\frac{1}{2}$	Good cinder forge
Millfields—J. T. Sparrow & Co.		0	Did make good iron
Priestfields, New—W. Ward and Sons.	$\frac{5}{4}$	1	Very strong grey forge
Osier Bed—Osier Bed Iron Co.	4	1	Good grey forge
Stow Heath—W. and J. S. Sparrow.	3	$1\frac{1}{4}$	Grey forge
Willenhall — Fletcher, Solly, and Urwick.	3	$2\frac{3}{4}$	Strong grey forge
BILSTON.			
Bilston Brook—Bilston Brook Furnace Co.	3	2	Cinder melters and cinder forge
Herbert's Park—D. Jones and Sons.	1	1	Good forge
Barbor's Field—Fowler, Holcroft and Hughes	2	$1\frac{1}{2}$	First-class grey forge
Caponfield — J. Bagnall and Sons.	3	2	First-class grey forge
Spring Vale—A. Hickman	3	$2\frac{3}{4}$	Mine grey forge

Name of Works and Owners.	Built.	In Blast.	Kind and Quality of Iron made.
BILSTON—*cont.*			
Deepfields—Deepfields Iron Co.	3	2	Cinder forge and melter
Coseley—J. and T. Turley	2	$1\frac{3}{4}$	Good mine forge
Priorfields—H. B. Whitehouse	3	3	Best melters
Stonefield—Stonefield Iron Co.	1	1	Cinder grey forge
Bradley—G. B. Thorneycroft and Co.	2	2	First-class grey forge
Bouvereux—Holcroft and Co. .	2	1	Good mine iron
WEDNESBURY.			
Rough Hay — Addenbrooke, Smith, and Pidcock.	3	2	First-class grey forge
Old Park—Patent Shaft and Axletree Co.	3	2	Very good grey
Broadwaters—S. Groucutt and Sons.	3	2	Grey forge, various
Darlaston—Darlaston Iron and Steel Co.	3	$2\frac{1}{2}$	Good grey forge
Moxley—David Rose	2	1	Good forge iron
TIPTON.			
Wednesbury Oak—P. Williams and Sons.	3	2	First-class cold blast mine iron
Willingsworth Iron Co.— Messrs. Pearson and Kendrick	2	2	Good forge iron
Tipton—J. and T. Turley	2	0	No return
Tipton Green—W. Roberts and Co.	4	$3\frac{3}{4}$	Good melters and cinder forge
Coneygree—Earl of Dudley .	3	3	Best in district
Park Lane—J. Colbourn and Sons.	2	2	Good mine forge
Horseley—J. Colbourn and Sons	2	2	Good mine forge
Groveland—G. Hickman .	2	2	Cinder forg
Hange Tividale—Round Bros.	2	2	First-class mine forge
Dudley Port—George Vernon	3	0	Cinder forge and melters
Do. J. and C. Onions .	1	1	Cinder forge and melters
WEST BROMWICH AND OLDBURY.			
Gold's Hill—J. Bagnall and Sons.	3	1	First-class mine

Name of Works and Owners.	Built.	In Blast.	Kind and Quality of Iron made.
WEST BROMWICH AND OLDBURY—*cont.*			
Union—P. Williams and Co. .	3	$1\frac{1}{2}$	First-class mine
Stour Valley—J. and C. Onions	2	2	Cinder melters and forge
Crookhay—H. and O. Firmstone.	4	2	Forge iron
Cape Smethwick . . .	1	0	
WALSALL.			
Roughwood—Williams Bros. .	2	0	
Hatherton—G. & R. Thomas .	2	$1\frac{2}{3}$	Make good mine iron
Birchills, New — J. Brayford (dead).	4	0	
Bentley—Chillington Iron Co.	2	2	Good mine
Pelsall—B. Bloomer and Son .	2	2	Good mine
Green Lanes—Jno. Jones and Sons., Walsall Iron Co.	2	$1\frac{1}{4}$	
Dixon's Green—Walsall Iron Co.	1	1	Cinder
Oldbury—J. & S. Onions .	4	0	
WEST OF DUDLEY.			
Corngreaves—New British Iron Co.	6	4	First-class quality
Withymoor—W. H. Dawes .	2	2	Very good mine. Make best hydrate iron
Netherton — N. Hingley and Sons.	4	3	Good iron
Windmill End—J. H. Pearson, Sir H. St. Paul	3	2	Good melter
New Level—Earl of Dudley .	4	3	The very best pigs
Netherton, New—M. and W. Grazebrook.	2	2	Quite the best cold blast pigs, made for sale. A noted speciality
Woodside—Cochrane and Co. .	3	$2\frac{3}{4}$	Good melters
Old Level—Hall, Holcroft and Pearson, Henry Hall	3	2	Good forge iron
Shutt End—J. Bradley and Co.	4	3	High-class forge
Corbyn's Hall, New—Corbyn's Hall Iron Co.	4	1	Forge iron, good

Name of Works and Owners.	Built.	In Blast.	Kinds and Quality of Iron made.
WEST OF DUDLEY—*cont.*			
Corbyn's Hall—Executors of W. Matthews.	4	2	Good mine forge
The Leys—W. and G. Firmstone	3	2	Strong mine forge
Parkhead—Evers and Martin .	2	2	Very good mine forge
Old Hill—David Rose . .	2	1	Good mine forge
Buffery—Jno. Jones and Sons .	1	1	Forge iron
Total	171	110	
NORTH STAFFORDSHIRE.			
Biddulph Valley — Robert Heath and Son	4	$3\frac{1}{2}$	Best mine forge
Norton—Robert Heath and Son	4	4	Strong mine forge
Clough Hall—Kinnersly and Co.	4	4	Good mine forge
Fenton Park — Fenton Park Iron Co.	2	0	Mine forge
Goldendale—Williamson Brothers	4	3	Good mine forge
Lane End—Thomas Goddard and Son	3	3	No. 3 and 4 melters
Shelton—Earl Granville . .	8	7	Good mine forge
Silverdale—Stanier and Co. ⎱ Apedale—Stanier and Co. ⎰	8	$5\frac{1}{2}$	⎰ Good mine forge ⎱ Good mine forge
Talke—New North Staffordshire Iron Co.	2	$1\frac{1}{2}$	Good mine forge
Total	39	$31\frac{1}{2}$	

FURNACES BUILDING.

Messrs. Goddard are about to erect one new furnace at Lane End.

Name of Works and Owners.	Built.	In Blast.	Kinds and Quality of Iron made.
DERBYSHIRE.			
Alfreton—J. Oakes and Co. .	3	$2\frac{1}{4}$	Grey forge
Butterley ⎱ Codnor ⎰ The Butterley Co.	7	5	Grey forge
Clay Cross—The Clay Cross Co.	3	2	Grey forge
Denby—W. H. & George Dawes	4	3	Grey forge

Name of Works and Owners.	Built.	In Blast.	Kinds and Quality of Iron made.
DERBYSHIRE—*cont.*			
Morley Park—C. C. Disney .	2	1½	Grey forge
Newbold—Newbold Iron Co. .	1	1	Grey forge
Oakerthorpe — Oakerthorpe Iron and Coal Co., Limited	2	0	Grey forge
Renishaw—Appleby and Co. .	4	2	Grey forge
Sheepbridge — Sheepbridge Coal and Iron Co., Limited	5	5	Grey forge
Stanton—The Stanton Iron Works Co.	5	5	Grey forge good
Staveley—Staveley Iron and Coal Co., Limited	7	6¼	Grey forge and melters
West Hallam—H. B. White-house & Sons	3	2	Grey forge very good
Wingerworth — Wingerworth Iron Co.	3	3	Grey forge good
Total	49	38	
SHROPSHIRE.			
Dark Lane ⎱ Leighton Green- Hinkshay ⎰ fell	} 4	3	Grey forge
Castle ⎱ Lawley ⎰ Coalbrookdale Lightmoor ⎰ Co.	{ 2 1 2	1 0 2	} Best grey forge
Ketley—Ketley Co. . .	1	1	Best grey forge very good
Lodge Wood ⎱ Lilleshall Iron Prior's Lee ⎰ Co.	} 9	8	{ Very best grey forge and melters
Madeley Wood—Madeley Wood Co.	3	3	Highest class grey forge and melters
Madeley Court—W. O. Foster	3	2	First-class grey forge
Old Park—Old Park Iron Co.	4	2	Grey forge
Total	29	22	
YORKSHIRE—WEST RIDING.			
Beeston Manor—A. Harding and Co.	2	1	First-class forge
Bowling—Bowling Iron Co. .	6	5	Highest class mel-ters. Grey forge

Name of Works and Owners.	Built.	In Blast.	Kinds and Quality of Iron made.
YORKSHIRE—WEST RIDING —*cont.*			
Elsecar—W. H. and George Dawes.	4		Good grey forge and melters
Milton—W. H. and George Dawes.	2		Good grey forge
Farnley—The Farnley Iron Co.	4		High class grey forge
Holmes—Parkgate Iron Co., Lim. and Parkgate	5		Good grey forge
Low Moor, Bradford—Hird, Dawson, and Hardy.	8	6	Highest class. Most valued iron in the kingdom
Thorncliffe, Chapeltown—Newton, Chambers, and Co.	2	1	Grey forge
Worsborough — Worsborough Iron Co.	1	0	
New York, Leeds—R. and W. Garside's Trustees.	2	2	Good iron
Hepworth—Hepworth Iron Co.	2	0	
Brightside—Cooke and Co. .	2	2	Good grey forge
West Ardsley — West Yorkshire Iron and Coal Co.	5	4	Good grey forge
Total	45	32	
NORTH-WEST OF ENGLAND.			
Duddon—Harrison Ainslie and Co.	1	1	Charcoal
Lonsdale Iron Co. . . .	2	1	Bessemer iron
Moss Bay—Hematite Iron Co.	2	1	Bessemer iron
Wigan—Wigan Coal and Iron Co., Lim.	10	9	Good *semi*-hematite forge
Ditton—Ditton Brook Iron Co., Lim.	6	4	Good hematite forge
Carnforth — Carnforth Hematite Iron Co., Lim.	5	5	Good hematite forge
Barrow—Barrow Hematite Steel Co., Lim	14	12	Noted Bessemer and strongest hematite grey forge and melting iron [1]
Cleator—Whitehaven Hematite Iron Co.	6	5	Gray Bessemer iron

[1] The very best No. 1 and 2 likewise made here.

Name of Works and Owners.	Built.	In Blast.	Kinds and Quality of Iron made.
NORTH-WEST OF ENGLAND *—cont.*			
Harrington—Bain, Blair, and Paterson	4	4	Good hematite forge iron and Bessemer
Workington—Workington Iron Co.	6	$4\frac{1}{4}$	High class Bessemer
West Cumberland—Hematite Iron Co., Lim.	5	4	Bessemer iron
Millom—Cumberland Iron Mining and Smelting Co., Lim.`	6	5	Best Bessemer iron
Askam—Furness Iron and Steel Co., Lim.	3	$2\frac{3}{4}$	Bessemer iron
Maryport—Gilmour and Co.	6	6	Bessemer iron
Solway—Solway Iron Co.	4	3	Bessemer iron
Furness—Furness Iron Co. .	2	0	
Total	8^2	67	

FURNACES BUILDING.

The Moss Bay Hæmatite Iron Co. are building two furnaces.

The Ditton Iron Co. are building two furnaces.

Blair and Co. are building two new furnaces at Pacton.

New North of England Iron Co. are building two new furnaces at Workington.

The Maryport Hæmatite Iron Co. are building one new furnace.

The Barrow Rolling Mills Co. are building two new furnaces.

The Lonsdale Iron Co. are building two new furnaces.

Name of Works and Owners.	Built.	In Blast.	Kinds and Quality of Iron made.
CLEVELAND.			
MIDDLESBOROUGH.			
Lackenby—Lackenby Iron Co.	3	2	Grey forge and melters
Eston — Bolckow, Vaughan, and Co.	7	7	Grey forge and melters
South Bank—T. Vaughan and Co.	9	$7\frac{1}{2}$	No. 1 and grey forge
Clay Lane—T. Vaughan and Co.	6	6	No. 1 and grey forge
Cargo Fleet—Swan, Coates, and Co.	4	4	Forge and melters
Normanby—Jones, Dunning, and Co.	3	3	Grey forge

Name of Works and Owners.	Built.	In Blast.	Kinds and Quality of Iron made.
MIDDLESBOROUGH—*cont.*			
Ormesby—Cochrane, and Co. .	4	3	Grey forge
Tees—Gilkes, Wilson, Pease, and Co.	5	4¾	No. 1 and grey forge
Middlesborough — Bolckow, Vaughan, and Co.	3	3	Grey forge
Tees Side—Hopkins, Gilkes, and Co.	4	4	No. 1 and grey forge
Linthorpe—Lloyd and Co. .	6	6	No. 1 and No. 3
Acklam—Stevenson, Jaques, and Co.	4	4	Grey forge
Ayresome—Gjers, Mills, and Co.	4	4	No. 1 and No. 3
Newport—B. Samuelson and Co.	8	7½	Grey forge
Clarence—Bell Brothers . .	8	8	Grey forge
Felling Gateshead—H. L. Pattinson and Co.	2	0	
STOCKTON-ON-TEES.			
Norton—Norton Iron Co. .	6	3	Best Bessemer and grey forge good
Norwegian—Titanic Iron Co.	2	2	Grey forge good
Thornaby—W. Whitwell and Co.	5	3	Grey forge
Stockton—Stockton Iron Co. .		3	Grey forge
Carlton—Industrial Iron Co. .	8	2	Grey forge
WHITBY.			
Grosmont—C. and T. Bagnall, Jun.	2	2	Good grey forge and melters
Glaisdale—S. Cleveland Iron Co.	3	3	No. 1 and No. 3
Total	104	91¾	
NORTH-EAST OF ENGLAND.			
DARLINGTON.			
Middleton—Geo. Wythes and Co.	3	3	Melters
South Durham—South Durham Iron Co.	3	3	Grey forge

Name of Works and Owners.	Built.	In Blast.	Kinds and Quality of Iron made.
FERRY HILL.			
Ferry Hill—Rosedale and Ferry Hill Iron Co.	8	8	No. 1 and good grey forge
Hareshaw—Hareshaw Iron Co.	2	0	
BISHOP AUCKLAND.			
Witton Park — Bolckow, Vaughan, and Co.	5	5	Grey forge
Tow Law—Weardale Iron Co.	4	3	Best in district made here. Bessemer iron
CONSETT.			
Consett—Consett Iron Co. .	12	6	Grey forge good
CHESTER-LE-STREET.			
Birtley—Birtley Iron Co. .	3	0	
WASHINGTON.			
Wear—Bell Brothers . .	1	1	No. 1 and No. 3 and grey forge
LEAMINGTON-ON-TYNE.			
Leamington-on-Tyne—Bulmer and Co.	2	0	
NEWCASTLE-ON-TYNE.			
Jarrow—Palmer's Shipbuilding and Iron Co.	4	4	Grey forge
Elswick—Sir Walter G. Armstrong and Co.	2	2	Melters good
Walker—Losh, Wilson, and Bell	2	2	Melters
Seaham—Watson, Kipling and Co.	1	1	Best Bessemer
Wylam—Bell Brothers . .	1	0	
Total	53	38	

FURNACES BUILDING.

The Rosedale and Ferry Hill Co. are building two furnaces.

Bolckow, Vaughan, and Co. are building one new furnace at Eston.

Whitwell and Co. are building two new furnaces.

Downey and Co. are building two new furnaces at Coatham Iron Works.

Teesbridge Co. are building two new furnaces.

Robson, Maynard, and Co. are building two new furnaces at Redcar.

T. Richardson and Co. are building three new furnaces at West Hartlepool.

Messrs. Watson, Kipling, and Co. are building one new furnace.

Messrs. Hopkins, Gilkes, and Co. are building three new furnaces.

The Lackenby Iron Co. are building one new furnace.

Cochrane and Co. are building two new furnaces.

Gjers, Mills, and Co. are building one new furnace.

B. Samuelson and Co. are building one new furnace.

Consett Iron Co., Limited, are building two new furnaces.

North of England Iron Co. are building one new furnace.

Name of Works and Owners.	Built.	In Blast.	Kinds and Quality of Iron made.
GLOUCESTERSHIRE.			
Cinderford—Henry Crawshay & Co.	4	3	Best grey forge
Oakwood—Ebbw Vale Iron Co.	1	0	Grey forge
Park End, Lydney—Forest of Dean Iron Co.	3	2	Best grey forge
Soudley, Newnham — Goold Brothers	2	1	Best grey forge
WILTSHIRE.			
Westbury—Westbury Iron Co., Limited	4	3	Grey forge
Seend—Malcolm and Co. .	3	2	Grey forge
SOMERSETSHIRE.			
Ashton Vale—Ashton Vale Iron Co., Limited	1	1	Good grey forge
HAMPSHIRE.			
Warsash—Harrison Ainslie and Co.	1	0	
Total, Gloucestershire, Wiltshire, Somersetshire and Hampshire	19	12	

Name of Works and Owners.	Built.	In Blast.	Kinds and Quality of Iron made.
NORTHAMPTON AND LIN-COLNSHIRE.			
East End and Frithlingborough, Wellingborough—T. Butlin and Co.	4	4	Grey forge good
Heyford, Weedon—Geo. Pell .	3	2	Grey forge good
Stowe, Weedon—Castle Dyke's Co.	2	0	Made but very little
Finedon, Northampton—Glendon Iron Co., Checkland and Fisher	3	3	Grey forge
Trent (North Lincoln)—W. H. and G. Dawes	3	2	Grey forge
Frodingham — Frodingham Iron Co.	4	$3\frac{3}{4}$	Grey forge
North Lincolnshire — North Lincolnshire Iron Co.	2	1	Grey forge
Islip—Thrapstone C. H. Plevins	1	0	
Total	22	$15\frac{3}{4}$	.

FURNACES BUILDING.

The Frodingham Iron Co. are building a new furnace.
The Lancaster Iron Co. are building two new furnaces at Mostyn.

Name of Works and Owners.	Built.	In Blast.	Kinds and Quality of Iron made.
NORTH WALES.			
Brymbo—Brymbo Iron Co. .	3	2	Grey forge very good
Ffrwd—Sparrow and Poole . .	3	$1\frac{3}{4}$	Grey forge very good
Leeswood—Leeswood Iron Co., Limited	2	1	Grey forge
Plaskynaston—Buckley, Newton, and Co.	1	0	
Ruabon—New British Iron Co.	3	1	Grey forge very good
Mostyn—John Lancaster and Co.	2	2	Bessemer
Total North Wales .	14	$7\frac{3}{4}$	

Name of Works and Owners.	Built.	In Blast.	Kinds and Quality of Iron made.
MONMOUTHSHIRE AND SOUTH WALES.			
Abersychan—Ebbw Vale Co. .	6	2	Grey forge
Ebbw Vale— Ditto . .	3	2	Grey forge
Victoria— Ditto . .	3	2	Grey forge
Sirhowey Ditto . .	4	4	Grey forge
Pontypool— Ditto . .	3	3	Best grey forg
Nant-y-Glo—J. & C. Bailey .	6	6	Grey forge
Blaenavon—Blaenavon Iron Co.	10	8	Best grey forge
Beaufort	6	6	Grey forge
Blaina—Levick and Co. . .	2	0	Grey forge
Cwm Celyn— Ditto . .	3	2	Grey forge
Coalbrook Vale—Ditto . .	1	1	Grey forge
Cwm Brân—Patent Nut and Bolt Co.	2	2	Best grey forge
Rhymney 6, Bute 3—Rhymney Co.	9	7	Best grey forge
Tredegar—Tredegar Iron Co. .	9	7	Grey forge
Total	67	52	
GLAMORGAN.			
Aberaman—Powell Duffryn .	3	3	Grey forge
Abernant 2, Llwydcoed 3— Fothergill and Co.	5	3	Grey forge
Plymouth 3, Duffryn 4— Fothergill and Co.	10	4	Grey forge
Penydarran—Fothergill and Co.	7	0	Grey forge
Brynna—Fothergill and Co. .	1	0	Grey forge
Briton Ferry — Townsend, Wood, and Co.	2	2	Grey forge
Cwn Avon—Copper Miners .	6	3½	Good grey forge
Cyfarthfa 5, Ynisvach 4—R. T. Crawshay, Ynisfach	11	9	Grey forge
Cefn Cwsk—Mr. Talbot .	4	0	Grey forge
Dowlais—Guest and Co. .	17	16	Grey forge
Gadlys—Wayne and Co. .	4	2	Grey forge
Llynvi—Llynvi Iron Co. .	7	4	Good grey forge
Maesteg—Maesteg Iron Co. .	3	0	Grey forge
Pentyrch—Booker and Co. .	2	2	Grey forge
Pontardawe—Lewis and Sons .	1	0	Grey forge
Pontypridd—Francis Crawshaw	2	0	
Tondu—Brogden and Sons .	2	2	Good grey forge
Treforest—F. Crawshay .	3	0	Grey forge

Name of Works and Owners.	Built.	In Blast.	Kinds and Quality of Iron made.
GLAMORGAN—*cont.*		.	
Ystalyfera [1]—Budd and Co. .	11	6	Grey forge
Venalt [1]—W. Gregory . .	1	0	Grey forge
Total	102	56½	
BRECON.			
Yniscedwyn [1] — Yniscedwyn Iron Co.	2	2	First-class forge and melting
Onllwyn, Neath — Onllwyn Iron Co.	2	0	Grey forge
Hirwain—Hirwain Iron Co. .	4	0	Grey forge
Beaufort—J. and C. Bailey .	7	6	Grey forge
Clydach—Basil Jayne . .	4	0	Grey forge
Total	19	8	
CARMARTHENSHIRE.			
Bryn Amman [1]—Strick and Co.		3	Good grey forge
Gwendreath—D. Watney .		0	Grey forge
Trimsaran (Gwendreath)— .	8	0	Grey forge
PEMBROKE.			
Kilgetty [1]—Vickerman and Co.	2	0	
Total	10	3	

FURNACE BUILDING.

The Blaenavon Company are building one new furnace.

Name of Works and Owners.	Built.	In Blast.	Kinds and Quality of Iron made.
SCOTLAND.			
Gartsherrie — Messrs. Wm. Baird and Co.	16	12	No. 1 melting
Coltness—Coltness Iron Co. .	12	12	No. 1 melting
Summerlee—Wilson and Co. .	8	7	No. 1 melting
Langloan—Robert Addie .	8	7	No. 1 melting
Govan—W. Dixon . .	5	5	No. 1 melting and No. 3

[1] Use anthracite coal.

Name of Works and Owners.	Built.	In Blast.	Kind and Quality of Iron made.
SCOTLAND—*cont.*			
Calder—W. Dixon . .	8	6	No. 1 melting
Carnbroe—Merry and Cuning-hame.	6	6	No. 1 and No. 3
Shotts—Shotts Iron Co. .		4	No. 1 melting
Castle Hill—Shotts Iron Co. .		3	No. 1 melting
Wishaw—Wishaw Iron Co. .		2	No. 1 and No.
Calderbank — Monkland Iron and Steel Co.	4/6	5	No. 1 and No. 3
Chappelhall — Monkland Iron and Steel Co.	3	3	No. 1 and No. 3
Clyde—Colin, Dunlop, and Co.	6	5	No. 1 and No. 3
Clyde (Quarter) Colin, Dunlop, and Co.	4	4	No. 1 and No. 3
EAST COAST.			
Kinneil—George Wilson and Co.	4	3	No. 1 and No. 3
Almond—James Russel and Son		2	No. 1 and No. 3
Carron—Carron Iron Co. .		3	No. 1
Lochgelly—Lochgelly Iron Co.	3/4	2	No. 1 and No. 3
Lumphinans — Lumphinans Iron Co.	2/3	1	No. 1 and No. 3
Bridgeness—Henry Cadell .	2	1	No. 1 and No. 3
WEST COAST.			
Églinton—W. Baird and Co. .	7	7	No. 1 and No. 3
Lugar— Ditto . .	4	4	No. 1 and No. 3
Muirkirk— Ditto . .	3	3	No. 1 and No. 3
Portland— Ditto . .	6	4	No. 1 and No. 3
Dalmellington—Dalmellington Iron Co.	8	7	No. 1
Glengarnock — Glengarnock Iron Co.	9	7½	No. 1
Ardeer—Merry and Cunning-ham	5	4	
HADDINGTONSHIRE.			
Glaselsmuir — Westbank, C. and A. Christie	1	1	
Lorn—Harrison, Ainsley, and Co.	1	0	
Total . . .	155	130½	

ABSTRACT OF TABULAR STATEMENTS.

BLAST FURNACES.

Name of Works and Owners.	Built.	In Blast.	Kinds and Quality of Iron made.
Cleveland	104	$91\frac{3}{4}$	The kinds and
North-East of England . .	53	38	qualities are so va-
North-West of England . .	82	67	ried in most dis-
South Staffordshire .	171	110	tricts, that it is
North Staffordshire .	39	$31\frac{1}{2}$	difficult with ac-
Shropshire . . .	29	22	curacy to describe
Yorkshire—West Riding	45	32	them. We may,
Derbyshire	49	38	however, say that
Northampton and Lincoln .	22	$15\frac{3}{4}$	the best iron is
Gloucester, Wilts, &c. . .	19	12	made near Bradford
North Wales	14	$7\frac{3}{4}$	in Yorkshire, Staf-
South Wales and Monmouth .	198	$119\frac{1}{2}$	fordshire, Shrop-
Scotland	155	$130\frac{1}{2}$	shire and Lanca-
	—	—	shire.
Total	980	$715\frac{3}{4}$	

CHARCOAL FURNACES.

Newland furnace, converting into hot blast—charcoal—
Harrison, Ainslie, and Co. 1 0

Backbarrow furnace, cold blast — charcoal—Harrison,
Ainslie, and Co. 1

Duddon furnace, cold blast — charcoal — Harrison,
Ainslie, and Co. (Cumberland) 1 0

Bonaire furnace, cold blast — charcoal — Harrison,
Ainslie, and Co. (Lorn, Argyleshire) . . . 1

Bonaire furnace, cold blast—charcoal (Bilston, Staffordshire) 2 2

Warsash furnace, cold blast—charcoal—Harrison, Ainslie,
and Co. (Hampshire) 1

LIST OF ALL THE PUDDLING FURNACES IN THE UNITED KINGDOM.

₊ So many changes are continually being made by the introduction of Danks's, Casson's, and other Patent Furnaces, and such large numbers of Puddling Furnaces are standing from one cause or other, that it is quite impossible to make this List of Puddling Furnaces as perfect as we could wish.

SOUTH STAFFORDSHIRE.—BLACK COUNTRY.

BILSTON.	Number of Puddling Furnaces.	Number of Mills and Forges.
Bank Field } Bradley } S. Groucutt and Son . . .	45	8
Batman's Hill—W. Rose	11	3
Bilston Mill—W. and J. S. Sparrow	26	4
Bradley Tin Plate—Thompson, Hatton, and Co.	8	4
Bradley—Stephen Thompson . . .	9	2
Bradley, New—Gittings and Austin	8	1
Deepfields—Bilston Iron Co.	12	1
Bilston Bridge and Herbert's Park—David Jones and Sons..	29	4
Millfield— Standing	14	1
Britannia—Brereton and Co.	11	3
Regent—Regent Iron Co.	11	3
Stonefield—Bilston Iron Co.	20	5
Ettingshall—Morewood and Co. . . .	12	3
Ebenezer—Standing, H. Onions and Co.	10	1
Millfields, New—Adams and Co. . . .	12	4

WOLVERHAMPTON DISTRICT.

	Number of Puddling Furnaces.	Number of Mills and Forges.
Lanesfield—Brookes and Merriman .	19	3
Chillington, Capponfield, Bilston, Leebrook, } Chillington Iron Co., Lim. }	95	6
Cleveland, Monmoor—E. T. Wright .	25	8
Horseley Fields—Osier Bed Iron Co. .	26	6
Minerva—Isaac Jenks and Sons .	21	4
Shrubbery, Swan Gardens—G. B. Thorneycroft and Co.	74	12

WILLENHALL.	Number of Puddling Furnaces.	Number of Mills and Forges.
Monmoor Lane—H. Deakin . .	15	3

WEDNESBURY DISTRICT.

Bull's Bridge—Molineux and Co. . . .	10	2
Darlaston Green—Darlaston Iron and Steel Co., Lim.	} 42	8
King's Hill—Darlaston Iron and Steel Co., Lim.		
Victoria, Moxley—David Rose .	8	1
Albert, Moxley—David Rose	22	3
Monway—J. Marshall	10	3
Old Park—Patent Shaft and Axle Co. .	32 }	8
Brunswick— Ditto ditto .	54 }	
Moxley—Thomas Wells	22	3

WALSALL DISTRICT.

Pleck Walsall—Skelton and Yardley	18	2
Victoria—H. Mills and Sons . . .	10	2
Wedge's Mills, Cannock—W. Gilpin, sen., and Co.	8	2
Birchills—J. Bissell and Son	16	3
New Birchills—Birchills Hall Iron Co., Lim. .	12	2
Staffordshire—Bunch, Jones, and Co. . .	10	2
Pelsall—B. Bloomer and Son	40	7
Cyclops—E. Russell	22	3

WEST BROMWICH DISTRICT.

Gold's Hill } Imperial } John Bagnall and Sons Leabrook }	75	8
Gold's Hill, New—T. Davis. .	13	2
Albion—Albion Sheet Iron Co.	10	2
Atlas—E. Parkes . . .	15	3
Brick House—R. Williams and Son	8	2
Bromford—J. Dawes and Sons	69	7
Hall End—J. T. and W. E. Johnson	9	1
Crookhay—W. and G. Firmstone .	20	3
Witton's Lane—Roberts, Tonks, and Co.	5	1
Excelsior—T. P. Allen and Co. .	12	2

WEST BROMWICH—*cont.*	Number of Puddling Furnaces.	Number of Mills and Forges.
Eagle—Eagle Coal and Iron Co. . . .	20	3
Great Bridge—Iron and Steel Co. . . .	10	2
Victoria, Swan Village—D. Hipkins and Sons	8	1
Ridgacre—Whitehouse Brothers and Co.	20	2
Wellington—Allen and Holden .	13	2
Roway—E. Page and Sons	16	2
Spon Lane—Patent Nut and Bolt Co. . .	11	2
Providence—Bridge, Gill, and Bridge	16	2
Albion—Britannia Iron Co. . .	10	2
Albion—Lees, J. B. S., and C. .	12	2
Dunkirk—Jordan and Co. . .	4	1
Waterloo—J. T. and W. E. Johnson	23	3
Bush Farm—Bush Farm Iron Co. . . .	18	2
Bradford—Bradford Iron Co. . . .	9	1
Richmond Works—Maddock's and Downing	8	1
Bromford—John Dawes & Son .	69	{ 7 2 }

SMETHWICK.

Gun Barrel—W. Marshall	4	1
Anchor—Standing	7	1
Smethwick—J. Stone	3	1
Capetown—W. H. Brooks	5	2
Crown—J. Nicklin	11	3
London Works—London Works Co.	12	2
Grove—Nash and Co. . . .	10	1
District—District Iron Co.	20	3
Rabone Bridge—Rabone Bridge Iron Co.	10	2
Vulcan—Standing	8	1
Regent—Beard and Eberhard . . .	11	3

OLDBURY.

Brades—Hunt and Sons . . .	12	3
Britannia—Bright, Perry, and Gettings .	13	2
Eagle—F. Simpson and Co.	10	1

TIPTON.

Bloomfield Factory Tipton Green } W. Barrows and Son . .	100	10

TIPTON—*cont.*	Number of Puddling Furnaces.	Number of Mills and Forges.
Globe and Tividale—J. P. Haynes .	9	2
Groveland—G. Hickman	24	3
Great Bridge—Great Bridge Iron and Steel Co., Lim.	20	2
Gospel Oak—Gospel Oak Iron Co.	24	5
Sheepwash Lane—Stonehewer and Co. .	16	2
Summer Hill—W. Millington and Co	16	4
Church Lane—District Steel and Iron Co.	10	2
Hope Iron Co.	6	2
Toll End—E. and T. Bayley	7	2
Wednesbury Oak—P. Williams and Sons	32	5
Dudley Port	8	1

DUDLEY.

	Number of Puddling Furnaces.	Number of Mills and Forges.
Portfield—James Holcroft	18	3
Dudley Port—Plant and Fisher . . .	20	3
Netherton—Hingley and Sons .	42	3
Dixon's Green—Dixon's Green Iron Co.	11	1
Corbyn's Hall	40	3

BRIERLEY HILL.

	Number of Puddling Furnaces.	Number of Mills and Forges.
Brockmore Tinplate—Budd and Co.	88	3
Hart's Hill—Hingleys and Smith .	33	3
Level—H. Hall	18	2
Round Oak—Earl of Dudley . .	54	5
Brierley } Corngreaves } New British Iron Co.	64	8
The Lays—Brown and Freer . .	33	7
Cradley Forge—S. Evers and Sons	18	3
Swindon—E. P. and W. Baldwin . . .	12	2

STOURBRIDGE.

	Number of Puddling Furnaces.	Number of Mills and Forges.
Brettel Lane } Bretwell Hall } T. Webb and Sons . .	21	4
Stourbridge } Brierley } Shutt End } John Bradley and Co.	64	8
Whittington—J. Williams and Co.	9	2
Hyde—Lee and Bolton (2 Siemens Furnaces at work here)	20	5

KIDDERMINSTER.	Number of Puddling Furnaces.	Number of Mills and Forges.
Cookley—J. Knight and Co.	18	2
Broadwaters—Thompson, Hatton, and Co.	11	2
Wilden—E. P. and W. Baldwin . . .	7	2
Total	2,160	

NORTH STAFFORDSHIRE.

TUNSTALL DISTRICT.

Ravensdale—Robert Heath and Son	56	
Chesterton—Chesterton Mining Co., Lim, .	28	3

BIDDULPH.

Biddulph—Robert Heath and Son . .	43	

NORTON-IN-THE-MOORS.

Norton—Robert Heath and Son .	44	

KIDSGROVE.

Clough Hall—Kinnersly and Co. . . .	81	7
Wheelock—Wheelock Iron and Salt Co.	24	2

STOKE-UPON-TRENT.

Shelton—Shelton Bar Iron Company	97	7
Cliff Vale—J. Bull and Son . .	26	5
Berry Hill—W. Bowers	26	2

NEWCASTLE-UNDER-LYME.

Silverdale—Stanier and Company .	24 ⎫	5
Knutton— Ditto . . .	33 ⎭	
Total	480	

NORTH OF ENGLAND.

MIDDLESBROUGH.

Middlesbrough—Bolckow, Vaughan, and Co., Lim.	67	11
Tees Side—Hopkins, Gilkes, and Co., Limited	100	5

MIDDLESBOROUGH—*cont.*	Number of Puddling Furnaces.	Number of Mills and Forges.
Newport—Fox, Head, and Company . .	42	4
Imperial—Jackson, Gill, and Company, Limited	32	2
West Marsh—West Marsh Iron Company .	20	2
Britannia—Britannia Iron Company, Limited	120	3
Ayrton—Jones, Brothers, and Company	23	2

STOCKTON.

North Yorkshire—North Yorkshire Iron Company, Limited	59	4
Thornaby—W. Whitwell and Company . .	33	3
Richmond—R. Jaques and Company . .	6	1
Westbourne—J. Holdsworth and Company	22	2
Malleable—Stockton Malleable Iron Company	58	5
Rail Mill—Stockton Rail Mill Company	70	3
West Stockton—West Stockton Iron Company	33	3
Moor—Shaw, Johnson, and Reay . .	30	2
Bowesfield—Bowesfield Iron Company .	30	2
Carlton—North of England Iron and Coal Company, Limited (Danks's Pat.)	8	

DARLINGTON.

Albert Hill ⎱ —Darlington Iron Company, Lim. Springfield ⎰	198	8
Skerne—Skerne Iron Works Company, Lim.	58	4
Rise Carr—Fry, Ianson, and Company .	32	4
Whessoe—Thomas Vaughan	36	2

FERRY HILL.

Tudhoe—Weardale Iron and Coal Company, Lim.	56	5

BISHOP AUCKLAND.

Witton Park—Bolckow, Vaughan, and Co., Limited	101	11
Bishop Auckland—Thomas Vaughan . .	30	2

CONSETT.

Consett—Consett Iron Company, Limited .	151	10

. . . T

	Number of Puddling Furnaces.	Number of Mills and Forges.
CHESTER-LE-STREET.		
Birtley—Birtley Iron Company . . .	6	1
FENCE HOUSES.		
Britannia—Hopper, Radcliffe, and Company	42	3
SUNDERLAND.		
Monkwearmouth—S. Tyzack and Company	32	2
Wear—Oswald and Company . . .	80	4
South Hylton—Raine, Brothers . . .	13	3
HARTLEPOOL.		
Hartlepool—Hartlepool Malleable Iron Co.	32	2
West Hartlepool—T. Richardson and Sons	109	3
Stranton—Stranton Iron and Steel Co., Lim.	20	2
NEWCASTLE-UPON-TYNE.		
Walker—Losh, Wilson, and Bell .	57	4
JARROW-ON-TYNE.		
Jarrow—Palmer's Shipbuilding and Iron Company, Limited	70	6
Hive—J. Elliot	13	1
GATESHEAD.		
Gateshead—Hawks, Crawshay, and Sons	69	5
Park—J. Abbot and Company, Limited	38	2
Felling—Felling Coal and Iron Company, Lim.	23	2
Team—Thomas Abbot	20	2
Total . .	2,018	

NORTH WALES.

RUABON.

	Number of Puddling Furnaces.	Number of Mills and Forges.
Broughton Hall—Broughton Iron Company .	12	2
Llay—Standing	6	1
Ruabon—New British Iron Company	38	3
Wrexham Stansty Forge . .	6	1
Pontysyllta	10	1
Total . . .	72	

SOUTH WALES AND MONMOUTHSHIRE.

	Number of Puddling Furnaces.	Number of Mills and Forges.
LLANELLY.		
Amman—Amman Iron Company . .	8	1
CARDIFF.		
Aberdare and Abernant—Fothergill and Hankey	69	4
Aberamman—Standing	17	
Pentyrch and Merlin Griffith—T. W. Booker and Co.	14	2
Penydarren—Fothergill and Company	19	2
Taff Vale—Standing	14	2
Treforest—J. Evans and Company .	6	5
BRITON FERRY.		
Briton Ferry—Townsend, Wood, and Company	42	4
ABERGAVENNY.		
New Clydach—Basil Jayne . . .	12	1
PORT TALBOT.		
Cwm Avon and Taibach—Governor and Company of Copper Miners	30	5
BRIDGEND.		
Llynvi Vale—Llynvi Tondu and Ogmore Company, Limited	54	6
MERTHYR TYDVIL.		
Cyfarthfa and Ynisfach—Robert T. Crawshay	72	4
Dowlais—Dowlais Iron Company .	161	13
Gadlys—Gadlys Iron Company .	23	2
Plymouth—Fothergill and Hankey	68	5

TONDU.

Tondu—Llynvi Tondu and Ogmore Company, Lim. included above

SWANSEA.	Number of Puddling Furnaces.	Number of Mills and Forges.
Ystalyfera—Budd and Company . . .	42	16

NEWPORT.

Llanelly—Standing	10	1
Nantyglo—Nantyglo and Blaina Company, Lim.	66	5
Aberyschan—Ebbw Vale Iron Company, Lim.	56 ⎫	
Victoria ⎱ Ebbw Vale ⎰ Ditto ditto .	165 ⎬	9
Pontypool Ditto ditto	16 ⎭	
Blaina ⎱ Cwm Celyn ⎬ Nantyglo and Blaina Iron Company, Lim. . Coalbrookdale ⎰	52	4
Blaenavon—Blaenavon Iron Company .	89	8
Pontnewynydd—	26	2
Varteg and Golynos—Partridge and Jones	23	2
Rhymney—Rhymney Iron Company	93	108
Tredegar—Tredegar Iron Company	80	5
Oakfields—J. C. Hill and Company	23	2
Cwm Bran—Patent Nut and Bolt Co.	20	3

ABERDARE.

Hirwain—Standing	19	2
Total . . .	1,306	

SHROPSHIRE.

COALBROOKDALE AND OKENGATES DISTRICT.

Horsehay—Coalbrookdale Co. . .	42	2
Stirchley—Leighton and Grenfell . .	30	4
Ketley—Ketley Iron Co.	20	3
Trench—Shropshire Iron Co., Lim.	24	3
Wombridge—Wombridge	10	3
Hadley—Nettlefold and Chamberlain	10	2

COALBROOKDALE AND OKENGATES DISTRICT—*cont.*	Number of Puddling Furnaces.	Number of Mills and Forges.
Lawton—Bullivant and Co.	10	1
Heybridge—Heybridge Co.	10	1
Old Park—Old Park Iron Co., Lim. (standing)	30	3
Snedshill—Snedshill Bar Iron Company	40	5
Hollinswood—Eagle Iron Company, Limited	16	3
Total . . .	232	

SOMERSETSHIRE.

BRISTOL.

Bower Ashton—Joseph Tinn .	10	1
Bristol—George Tinn	3	1
Total .	13	

LANCASHIRE.

MANCHESTER.

Pendleton—W. Barningham	20	4
Ashbury—Ashbury Carriage and Iron Co.	17	2
Ashbury—Maybury, Matthews, and Co.	8	1
Wire Works—Richard Johnson and Nephew .	20	3
Rail Mill—Manchester, Sheffield, and Lincolnshire Railway Works	9	2
Staleybridge—John Summers . . .	6	1
Parkbridge—Hannah Lees and Sons	4	1
Oldham—Platt Brothers and Co. .	8	1

BOLTON.

Bolton—Bolton Iron and Steel Co.	6	5
Atlas Forge—Thomas Walmsley .	16	2

WARRINGTON.

Dallam—Dallam Forge Co., Lim. . . .	30	3
Bewsey—Wire Iron Co., Lim. . . .	43	6

Warrington—*cont.*	Number of Puddling Furnaces.	Number of Mills and Forges.
Whitecross Wire Works } Whitecross Wire Co., Lim. . . }	11	2
Ince Hall—Ince Hall Rolling Mills Co., Lim.	20	2

Preston.

Preston—North of England Carriage and Iron Co., Limited	35	3

Wigan.

Albion—Hall and Matthews	10	1
—— Dallam Forge Co.	31	2

Liverpool.

Mersey—Mersey Iron and Steel Co., Lim.	4	2
[1]Garston—Liverpool and Garston Iron and Steel Co., Lim.	46	3
Total	338	

NORTH-WEST OF ENGLAND.

Workington.

West Cumberland—W. Cumberland Iron and Steel Co., Lim.	40	3
Kirk Brothers and Co.	16	2
Moss Bay—Kirk and Valentine . . .	11	2
Marsh Side Iron Works—Joseph Price, Jun., and Co.	6	2

Maryport.

Ellen—Ellen Rolling Mills Co., Lim.	12	3
Total	85	

[1] Have six double puddling furnaces on Siemens' plan, reckoned twelve in return.

YORKSHIRE—WEST RIDING.

LEEDS.

	Number of Puddling Furnaces.	Number of Mills and Forges.
Albert—W. T. Coghlan and Drury .	24	3
Clarence—Taylor Brothers and Company	17	5
Farnley—Farnley Iron Co. . .	24	4
Kirkstall Forge—Kirkstall Forge Co.	24	3
Leeds—S. T. Cooper and Co. . . .	13	6
Monk Bridge—Monk Bridge Iron Co.	26	8
[1] Perseverance—J. Whitham and Son	40	4
Thornhill—Monk Bridge Iron Co.	12	3
Hunslet—Tyers, Middleton, and Co.	14	6
And 3 Bull furnaces.		

BRADFORD.

Bowling—Bowling Iron Co.	32	6
Low Moor—Hird, Dawson, and Hardy	40	7
Water Lane—James Perkins . . .	40	4

WAKEFIELD.

Calder Vale—Samuel Whitham .	30	4

NORMANTON.

Railway Iron Works—W. Thomson and Co.	25	2
Total	358	

SOUTH YORKSHIRE.

SHEFFIELD.

Atlas—John Brown and Co., Lim.. . .	72	17
Wortley—Andrews, Burrows, and Co.	11	4
Cyclops—C. Cammell and Co., Lim.	60	12
Elsecar and Milton—W. H. and George Dawes	61	10
And 6 Siemens' double furnaces.		

---

[1] Messrs. Whitham have 11 double machine puddling furnaces, equal to 36 hand furnaces.

ROTHERHAM.	Number of Puddling Furnaces.	Number of Mills and Forges.
Phœnix—Owens' Wheel and Tyre Co. . .	26	2
Parkgate—Parkgate Iron Co., Lim. .	90	
Midland—Midland Iron Co., Lim.	29	
Northfield—Northfield Co., Lim. .	32	
Rotherham—G. and J. Brown, Limited .	27	6
Total . . .	367	

DERBYSHIRE.
ALFRETON.

	Number of Puddling Furnaces.	Number of Mills and Forges.
Butterley—The Butterley Co. . . .	42	11

CHESTERFIELD.

Whittington—T. Firth and Sons . . .	18	3
Sheepbridge—Sheepbridge Coal and Iron Co., Lim.[1]	27	2

DERBY.

Railway—Eastwood, Swingler, and Company	14 ⎫	
Victoria—Eastwood, Swingler, and Company	25 ⎬	3
	⎭	
Total . .	126	

SCOTLAND.
GLASGOW.

Blochairn—Hannay and Sons . . .	50	4
Glasgow—Glasgow Iron Company	47	6
St. Rollox— Ditto . . .	13	2
Govan—William Dixon . .	40	5
Muirkirk—William Baird and Company	12	2
Parkhead—W. and J. Beardmore . . .	28	3

[1] 9 double furnaces worked by mechanical power, and equal to 27 ordinary furnaces.

COATBRIDGE.	Number of Puddling Furnaces.	Number of Mills and Forges.
Rochsolloch—Rochsolloch Iron Company	14	2
Clifton—Gray and Wyllie . . .	20	2
Coats—Thomas Jackson	27	2
Wishaw—John Williams and Company	12	2
Drumpeller—Henderson and Dimmack	16	2
North British—Thomas Ellis . .	12	3
Phœnix—John Spencer	22	1
Globe—A. and T. Miller	5	1
Coatbridge Tin Plate Works—Coatbridge Tin Plate Co.	11	2
Coatbridge Iron Works—Hugh Martin and Son	7	2
Monkland—Monkland Iron and Steel Company, Limited . . .	60	5

HOLYTOWN.		
Mossend—Mossend Iron Company . .	60	5
Clydesdale—Clydesdale Iron Company, Limited	10	2
Motherwell—Glasgow Iron Company	52	3
Dalziel—David Colville	15 .	2
Total . . .	565 .	

ABSTRACT LIST OF ALL THE PUDDLING FURNACES IN THE UNITED KINGDOM.

North of England	2,018
North-West of England . . .	84
Yorkshire—West Riding . . .	358
Yorkshire South	367
Derbyshire	126
South Staffordshire	2,160
North Staffordshire	480
Shropshire	230
Lancashire	338
Somersetshire	13
North Wales	70
South Wales and Monmouth .	1,306
Scotland	565
Total	8,115

LIST OF SMELTING AND METAL EXTRACTION COMPANIES IN THE UNITED KINGDOM.

LEAD.

Stock and Company, Penclawd, Swansea.
Sims, Willyams, Neville, and Company, Llanelly.
The Bury Port Smelting Company, Pembrey, Carmarthenshire.
Runcorn Smelting Company, Runcorn.
Thomas Somers, Bristol.
The Panther Lead Works, Bristol.
Sheldon, Bush, and P. S. Company, Redcliff Hill, Bristol.
Weston, Son, and Company, Bristol.
W. J. Cookson, and Company, Newcastle.
Locke, Blackett, and Company, Newcastle.
John Warwick, Newcastle.
Shield and Dinning, Haydon Bridge.
Howden Smelting Company, Newcastle.
Washington Chemical Company, Newcastle.
Enthoven and Sons, London.
Lock, Lancaster, and Company, London.
Pontifex and Wood, Farringdon Works, London.
Trustees Treffry's Estate, Par, Cornwall.
R. Michell and Son, Truro, Cornwall.
Peter Glover, Widness Lead Works, near Warrington.
Delafield White Lead Company, near Wrexham, Flintshire.
Adam Eyton, Llanerchymor, Holywell.
The Brymbo Company, Brymbo.
Walker, Parker, and Company, Dee Bank, Bagilt, and Newcastle.
Governor and Company of Lead Smelters, Nenthead, Alston Moor.
W. B. Beaumont, Allendale, Alston Moor.
Benjamin Bagshaw, Eyam, near Bakewell.
Barker and Rose, Alport, Bakewell, and Sheffield.
Joseph Wass and Company, Lea Lead Works, Matlock, Bath.
Snailbeach Lead Company, near Shrewsbury.
Pontesford Smelting Company.
Meerbrook Lead Works (Wass and Co.,) Meerbrook, Matlock, Bath.
Gibbs and Company, Bonsale Dale, Matlock.

LEAD—*cont.*

William Sperry, Via Gellia Lead Works, Cromford, Matlock, Bath.

Robert Howe Ashton, Castletown, Derbyshire.

J. Fairburn and Company, Middleton Dale and Bradwell.

T. Wilson and Company.

E. Backhouse, Darlington.

Greenside Mining Company, Penrith.

The Keld Head Mining Company, Wensleydale.

George York, Pateley Bridge.

Duke of Devonshire, Grassington.

The Duke of Buccleuch, Wanlock Head.

The Lead Hills Mining Company, Lead Hills.

The Hurst Mining Company, Hurst.

Lister, Robinson, and Company, Grinton Moor.

R. M. Jaques and Company, Arkengarthdale.

R. M. Jaques and Company, Old Gang.

The Blakethwaite Lead Company, Blakethwaite.

R. Milner and Company, Belde Hill.

The Swaledale Lead Company, West Swaledale.

Somersetshire:

    Charterhouse, Blagdon, Bristol.

    Mendip Mining Company.

    Waldegrave Lead Smelting Company Limited, Mendips, near Wells.

    East Harptree Lead Works, Limited, East Harptree, Bristol.

    St. Cuthbert Works, Mendips, near Wells.

The Mining Company of Ireland Limited, Dublin.

LIST OF ALL THE ZINC SMELTERS.

Vivian and Sons, Swansea.

William Marsden, Oldland Hall, near Bristol.

Kenrick and Son, Wynn Hall, Spelter Works, Ruabon.

Mines Royal Copper Company, Neath (ceased to smelt).

Charles Titterton, Phœnix Zinc Works, Warrington Junction.

J. H. Dillwyn, M.P., Swansea.

J. Collingborne, Spelter Works, Warmley, Bristol.

Joseph Thompson, Spelter Works, Carlisle.

ZINC SMELTERS—*cont.*

The Bagilt Smelting Company, Bagilt.

Ryland Brothers, Warrington.

T. Gilby.

Joseph Wethered, Bristol.

PYRITES PRECIPITATE COMPANIES.

Duncan McKechnie, St. Helen's.

The Widnes Metal Company, Widnes.

The Tharsis Sulphur and Copper Company, Widnes.

      ,,   ,,   ,,   Newcastle.

      ,,   ,,   ,,   Birmingham.

      ,,   ,,   ,,   Glasgow.

N. Mathieson and Company, Widnes.

The Runcorn Soap and Alkali Company, Runcorn.

Newton Heath Reduction Company, Manchester.

Muspratt Brothers and Huntley, Flint.

The Mostyn Copper Company, Mostyn.

Solomon Mease and Son, Newcastle.

William Russell and Company, Newcastle.

J. and W. Allan, Newcastle.

The Bede Metal Company, Newcastle.

W. Hunt and Sons, Leabrook, Wednesbury.

William Hunt and Company, Castleford.

H. Blair and Company, Kearnsley, Bolton, Lancashire.

William Haslam, Bolton, Lancashire.

Snell and Company, Runcorn.

ARSENIC.

Cornwall Arsenic Company, Hayle, Bissoe Bridge, Thomas Willis Field, Managing Partner.

LIST OF THE BEST HÆMATITE MES IN THE WHITEHAVEN DISTRICT, WHICH INCLUDES FRIZZINGTON.

No.			
1	John Stirling, Esq., Montreal Iron Ore Works	.	John Je ꞏꞏꞏ, Esq., Manager.
2	The Parkside Mining Co.	Moorkow, near Cleator M o..	Charles Fisher, Esq., Managing Director or Partner.
3	S. and J. Lindew	Frizzington	John Lindew, Esq., Managing ꞏꞏ or Partner.
		Gutterby and Rig Rigg, nr Cleator.	
4	The Salter and Eskett Park Mining Co., lim.	E ꞏꞏtt	Joseph C. Brown, Esq., Managing Director.
5	S. W. Smith, Esq.	Frizzington	Mr. E. G Hughes, Resdent ꞏꞏ
9	Fletcher and Hodgett	Frizzington	Alfred Hodgett, Esq., Managing Partner.
10	ꞏꞏos Clarke, Esq.	Birles, Frizzington	Jnes ꞏꞏ, Esq., Newton Heath, Manchester, ꞏꞏtner.
6	The Eskett Iron Ore G	Eskett	Joseph H. Robinson, Esq., Managing Director.
13	Fletchers, ꞏꞏ and G	Eskett	W. Miller, Esq., Managing Director.
19	Hodgson and G	ꞏꞏuse, Eskott	John Hodgson, Esq., Managing Director.
11	Right Hon. Lord Leconfield	Rig Rigg	James Davidson, Esq., Manager.
12	Bain and ꞏꞏrsur	ꞏꞏ mines, Eyre ꞏꞏm	Bain and Pate on.
16	Hannay an l Co.	Frizzington	H. Woolcock, Esq., Manager (sold).
18	The Lamplugh Mining Co.	ꞏꞏ	M. Stewart, Manager.
7	The Crossfield Iron Ore Co.	Crossfield	H. M. Macko ꞏꞏa, Esq., Managing Partner.
15	The ꞏꞏd Iron Mining and Smelting Co. lim.	Winder	Thomas ꞏꞏ, Esq., ꞏꞏg Director.
14	The Rig Rigg Iron Ore G	Rig Rigg	James ꞏꞏs, Esq., Managing Part n
8	The G ꞏꞏ Iron Ore G	Cleator	James Ai ꞏꞏ, Esq., Managing Partner.
17	The ꞏꞏy Iron Ore G	ꞏꞏy, Frizzington	James Robertson, Esq., Manager.

LIST OF WORKS HAVING BESSEMER CONVERTERS IN GREAT BRITAIN IN 1872.

No.	Name and Situation of Works	Number of Converters	Capacity of Converters	
			Tons.	Cwt.
1	Henry Bessemer and Co., Sheffield	2	3	0
		2	5	0
2	John Brown and Co., *Limited*, Sheffield	2	10	0
		4	6	0
3	Charles Cammell and Co., *Limited*, Sheffield	8	5	0
4	Weardale Iron Co., Towlaw	4	2½	0
5	The Glasgow Bessemer Steel Co., *Limited*, Atlas Works, Glasgow	2	3	0
6	Samuel Fox and Co., Stockbridge Works, Deepcar	2	5	0
		2	3	0
7	Lloyds, Foster, and Co., Old Park, Wednesbury	4	4	0
8	Bolton Iron and Steel Works, Bolton	4	5	0
9	London and North-Western Railway, Crewe	2	3	0
10	Lancashire Steel Co., Gorton	4	5	0
11	Mersey Steel and Iron Works, Liverpool	4	5	0
12	Manchester Steel and Railway Plant Co., Gibraltar Works, Newton Heath, Manchester	4	3	0
13	Barrow Hæmatite Steel Co., Barrow	18	6	0
14	The Dowlais Iron Co., Dowlais	6	5	0
15	Ebbw Vale Co., Ebbw Vale	7	5	0
16	Steel Ordnance Co., Limited, Greenwich	2	5	0
17	West Cumberland, Workington	4	7½	0
18	Phœnix Iron Co., Rotherham	2	3½	0
19	Carnforth Hæmatite Iron Co., *Limited*	2	0	0

LIST OF ALL THE TIN PLATE MANUFACTURERS. 1872.

No.	Name of Works	Name of Firm	Where situate
1	Aberdulais	JoshuaWilliams&Co.	Neath, Glamorganshire.
2	Abertillery	Philip S. Milliss	Newport, Monmouth-shire.
3	Beaufort or Lower Forest	BeaufortTinPlateCo.	Swansea, Glamorgan-shire.
4	Bradley	Thompson, Hatton, & Co.	Bilston, Staffordshire.
5	Broadwaters	Thompson, Hatton, & Co.	Kidderminster, Worces-tershire.
6	Brockmoor	Budd & Co.	Brierley Hill, Stafford-shire.
7	Caerleon	J. G. & A. Moggridge	Caerleon, Monmouth-shire.
8	Carmarthen	Thomas Lester & Co.	Carmarthen, Carmar-thenshire.

LIST OF ALL THE TIN PLATE MANUFACTURERS—*cont.*

No.	Name of Works	Name of Firm	Where situate
9	Coatbridge .	Coatbridge Tin Plate Co.	Coatbridge, Glasgow.
10	Cookley .	John Knight & Co. .	Kidderminster, Worcestershire.
11	Cwm Avon .	Governor & Co., Copper Miners	Taibach, Glamorganshire.
12	Cwmfelin .	Cwmfelin Tin Plate Co.	Swansea, Glamorganshire.
13	Dalen . .	Philips, Ninnes, &Co.	Llanelly, Carmarthenshire.
14	Derwent .	W. Griffiths & Co. .	Workington, Cumberland.
15	Garth . .	Garth Iron and Tin Plate Co.	Newport, Monmouthshire.
16	Gwendraeth .	J. Chivers & Son .	Kidwelly, Carmarthenshire.
17	Hendy . .	Edmd. Boughton & Co.	Llanelly, Carmarthenshire.
18	Horseley Fields	Osier Bed Iron Co. .	Wolverhampton, Staffordshire.
19	Horseley Fields or Wilden	E. P. & W. Baldwin	Ditto.
20	Landore .	Landore Tin Plate Co.	Swansea, Glamorganshire.
21	Lydbrook .	Richard Thomas & Co.	Ross, Herefordshire.
22	Llwydarth .	Llwydarth Tin Plate Co.	Bridgend, Glamorganshire.
23	Macken .	Macken Iron and Tin Plate Co.	Newport, Monmouthshire.
24	Melyn . .	Leach, Flower, & Co.	Neath, Carmarthens.
25	Pentyrch and MelinGriffith	T. W. Booker & Co.	Cardiff, Glamorganshire.
26	Pontardawe .	W. Gilbertson & Co.	Swansea, Glamorganshire.
27	Pontnewydd .	B. Conway & Co. .	Newport, Monmouthshire.
28	Ponthur .	John Jenkins & Co.	Caerleon, Monmouthshire.
29	Pontrhydyrnn	Conway Brothers .	Newport, Monmouthshire.
30	Pontymister .	Banks & Co. . .	Ditto.
31	The Old Castle	The Old Castle Iron and Tin Plate Co., *Limited*	Llanelly, Carmarthenshire.
32	Tividale .	Budd & Co. . .	Tipton, Staffordshire.
33	Upper Forest	Edward Bagot & Co.	Swansea.
34	Ynispeullwch	Ynispeullwch Tin Plate Co.	Swansea, Glamorganshire.

PRICES OF MERCHANT BAR IRON AT LIVERPOOL FROM 1806 TO 1872.

Date	Highest Price in the Year			Lowest Price in the Year			Average		
A.D.	£	s.	d.	£	s.	d.	£	s.	d.
1806	17	10	0	16	0	0	16	15	0
1807	17	0	0	15	0	0	16	0	0
1808	15	0	0	14	10	0	14	15	0
1809	16	0	0	14	0	0	15	0	0
1810	15	0	0	14	0	0	14	10	0
1811	15	0	0	14	0	0	14	10	0
1812	14	5	0	12	15	0	13	10	0
1813	13	0	0	12	0	0	12	10	0
1814	14	0	0	12	0	0	13	0	0
1815	13	10	0	11	0	0	12	5	0
1816	11	10	0	8	15	0	10	2	6
1817	13	0	0	8	10	0	10	15	0
1818	13	0	0	10	0	0	11	10	0
1819	12	10	0	11	0	0	11	15	0
1820	11	0	0	9	10	0	10	5	0
1821	9	10	0	8	15	0	9	2	6
1822	8	10	0	8	0	0	8	5	0
1823	8	10	0	8	0	0	8	5	0
1824	13	0	0	8	10	0	10	15	0
1825	15	0	0	11	10	0	13	5	0
1826	11	10	0	9	10	0	10	10	0
1827	10	0	0	8	15	0	9	7	6
1828	9	10	0	8	5	0	8	17	6
1829	8	5	0	7	0	0	7	12	6
1830	6	15	0	6	5	0	6	10	0
1831	6	10	0	5	15	0	6	2	6
1832	6	5	0	5	10	0	5	15	0
1833	7	15	0	5	15	0	6	17	6
1834	8	5	0	7	10	0	7	7	6
1835	8	5	0	7	10	0	7	7	6
1836	11	10	0	8	0	0	9	15	0
1837	10	10	0	7	5	0	8	17	6
1838	9	10	0	10	0	0	9	15	0
1839	10	5	0	9	10	0	9	17	6
1842	9	10	0	7	10	0	8	10	0
1843	6	15	0	5	10	0	6	2	6
1844	6	5	0	4	15	0	5	10	0
1845	10	5	0	6	5	0	8	5	0
1846	9	10	0	8	10	0	9	0	0
1847	9	10	0	8	15	0	9	2	6
1848	9	0	0	6	0	0	7	10	0
1849	6	10	0	5	0	0	5	15	0
1850	6	15	0	5	10	0	6	2	6
1851	5	0	0	4	10	0	4	15	0
1852	9	0	0	4	10	0	6	15	0
1853	9	10	0	7	10	0	8	10	0
1854	9	15	0	9	0	0	9	7	6

PRICES OF MERCHANT BAR IRON AT LIVERPOOL FROM 1806 TO 1872—*continued*.

Date	Highest Price in the Year			Lowest Price in the Year			Average		
A.D.	£	s.	d.	£	s.	d.	£	s.	d.
1855	9	0	0	7	0	0		0	0
1856	8	10	0	8	0	0		5	
1857	8	5	0	7	5	0		15	
1858	7	5	0	6	10	0		17	
1859	6	15	0	6	5	0		0	
1860	6	10	0	6	5	0		7	
1863	7	15	0	6	0	0		17	
1864	7	15	0	7	5	0		0	
1865	7	15	0	7	0	0		7	
1866	7	5	0	6	5	0		15	
1867	7	5	0	6	0	0		2	
1868	6	5	0	6	0	0		2	
1869	6	15	0	6	0	0		7	
1870	7	0	0	6	0	0		10	
1871	6	15	0	6	15	0	6	15	0
1872	16	0	0	11	0	0	13	12	6
1873	16	0	0	14	0	0	This being September, we cannot give average until January, 1874.		

COMPENDIUM OF MECHANICS.

WEIGHTS AND MEASURES.

IMPERIAL STANDARD WEIGHTS AND MEASURES.

STANDARD YARD.

The Standard Yard, when compared with a Pendulum vibrating seconds of mean time in the latitude of London, in a vacuum at the level of the sea, is in the proportion as 36 to 39 inches, and 1393 ten thousandth parts of an inch.

The Rood of Land shall contain 1210 square yards, an Acre, 4840 square yards, or 160 square perches, poles, or rods.

STANDARD POUND.

A Cubic Inch of distilled water, weighed in air by brass weights, at the temperature of 62 degrees Fahrenheit, the barometer being at 30 inches, is equal to 252 grains and 458 thousandth parts of a grain, of which the Standard Troy Pound shall contain 5760.

STANDARD GALLON.

The Standard Gallon shall contain 10 Pounds Avoirdupois weight of distilled water weighed in air, at the temperature of 62° Fahrenheit, the barometer being at 30 inches. It shall contain $277\frac{1}{4}$ cubic inches.

STANDARD FOR HEAPED MEASURE.

The Standard Bushel, for Heaped Measure, shall contain 80 pounds Avoirdupois weight. It shall be $19\frac{1}{2}$ inches diameter, and 7·4272 inches deep inside, with a plain and even bottom, and contain 2218·192 cubic inches.

1. MEASURES OF LENGTH.

12 inches	= 1 foot.
3 feet	= 1 yard.
5½ yards	= 1 pole or rod.
40 poles	= 1 furlong.
8 furlongs, 1760 yards, or 5280 feet	= 1 mile.

3 miles = 1 league, marked lea.	$\frac{11}{12}$ mile = 1 Italian mile.
2¾ do. = 1 French league.	¾ do. = 1 Russian verst.
3⅔ do. = 1 Spanish league.	$1\frac{3}{22}$ do. = 1 Scotch mile.
4 do. = 1 German mile.	$1\frac{3}{11}$ do. = 1 Irish mile.
3¼ do. = 1 Dutch mile.	$69\frac{1}{13}$ miles nearly = 1 Deg., marked °

VARIOUS FRENCH MEASURES OF FREQUENT REFERENCE.

A point, is equal to	·0148025 English inches.
A line	·088815 ,,
A millemetre	·039371 ,,
A centimetre	·39371
An inch (pouce)	1·06578
A decimetre	3·9371
A foot	12·78933 ,,
A metre	39·371 ,, or 3·2809 English ft.
A toise (fathom)	6·394 English feet.
A league	14591·1 ,, or 4863·7 English yards.
A square inch	1·13582 English square inches.
A cubic inch	1·21063 ,, cubic ,,
A cubic metre	35·316 ,, cubic feet.

2. MEASURES OF SURFACE, OR SQUARE MEASURE.

144 square inches	= 1 square foot.
9 square feet	= 1 square yard.
30¼ square yards	= 1 square pole.
40 square poles	= 1 square rood.
4 roods, or 4840 square yards	= 1 square acre.
640 square acres	= 1 square mile.
1089 Scotch acres	= 1369 Eng. acres.

3. MEASURES OF SOLIDITY OR CUBIC MEASURE.

1728 cubic inches	= 1 cubic foot.
27 cubic feet	= 1 cubic yard.
166⅜ cubic yards	= 1 cubic pole.
64,000 cubic poles	= 1 cubic furlong.
512 cubic furlongs	= 1 cubic mile.

4. MEASURES OF CAPACITY.

LIQUIDS.

8·665 cubic inches = $\frac{5}{16}$ ℔. of water	.	.	. = 1 gill.[1]
4 gills = 34·659 do. = 1¼ ℔. do. .	.	.	. = 1 pint.
2 pints = 69·318 do. = 2½ ℔s. do. .	.	.	. = 1 quart.
4 quarts = 277¼ do. = 10 ℔s. do. .	.	.	. = 1 gallon.

GRAIN, FRUIT, ETC.

2 gallons	.	.	. = 1 peck.
4 pecks, or 2218·192 cubic inches		.	= 1 bushel.
8 bushels	.	.	. = 1 quarter.
5 quarters	.	.	. = 1 load.

5. MEASURES OF WEIGHT.

TROY.

24 grains	.	.	.	. = 1 pennyweight.
20 pennyweights	.	.	. = 1 ounce.
12 ounces			. = 1 pound.

AVOIRDUPOIS.

27·34375 troy grains	.	.	.	. = 1 dram.
16 drams	.	.	.	. = 1 ounce.
16 ounces	.	.	.	. = 1 pound.
14 pounds				. = 1 stone.
2 stones	.	.	.	. = 1 quarter.
4 quarters, or 112℔s. .		.	.	. = 1 cwt.
20 cwt.	.	.	.	. = 1 ton.

BRITISH SPECIAL MEASURES.

1. LINEAL MEASURES FOR LAND.

7·92 inches	.	.	.	. = 1 link.
100 links, or 22 yards		.	.	. = 1 chain.
80 chains		.	.	. = 1 mile.
69·121 miles .		.	.	. = 1 geog. degree.

---

[1] Called also, in the north of England, a Jack or Noggin. In some counties, the *Half-pint* is erroneously termed a *Gill.*

2. SQUARE MEASURES FOR LAND.

62·7264 square inches	= 1 square link.
10,000 square links	= 1 square chain.
10 square chains	= 1 square acre.

3. NAUTICAL MEASURES.

6082·66 feet	= 1 nautical mile.
3 miles	= 1 league.
20 leagues	= 1 degree.
360 degrees	= the earth's circumference.

MISCELLANEOUS SPECIAL MEASURES.

6 lineal feet	= 1 fathom.
100 square feet	= 1 square of flooring.
272 square ft. at 14 inches in thickness	= 1 rod of brickwork.
600 square feet of inch boards	= 1 load.
40 cubic feet of round timber } 50 cubic feet of hewn timber }	= 1 ton or load.
40 cubic feet	= 1 ton of shipping.
120 deals	= 1 hundred.
120 nails	= 1 hundred.
1200 do.	= 1 thousand.
500 bricks	= 1 load.
32 bushels of lime	1 do.
36 do. sand	= 1 do.
22 cwt.	= 1 fodder of lead (Stockton).
21 do.	= 1 do. do. (Newcastle).
19½ do.	= 1 do. do. (London).
108 cubic feet	= 1 stack of wood.
42 gallons	= 1 tierce ⎫
63 do.	= 1 hogshead ⎪ old
84 do.	= 1 puncheon ⎬ wine
126 do.	= 1 pipe ⎪ measure.
252 do.	= 1 tun ⎭
36 do.	= 1 barrel ⎫
54 do.	= 1 hogshead ⎪ old ale
72 do.	= 1 puncheon ⎬ measure.
108 do.	= 1 butt ⎭
36 bushels, or 28 cwt	= 1 chaldron of coals (London).
53 cwt.	= 1 do. do. (Newcastle).
88 lbs.	1 bushel of coal.
56 do.	1 do. flour or salt.
60 do.	= 1 do. wheat.
47 do.	= 1 do. barley.
38 do.	= 1 do. oats.

MISCELLANEOUS SPECIAL MEASURES.—*Continued*.

1 gallon of sea water   .   = 10·32 ℔s. avoirdupois.
1   „   oil   .   .   =   9·32   „
1   „   proof spirits   .   =   9·3   „

The old ale gallon contained 282 cubic inches; and
the old wine gallon 231.

The French litre, or standard measure of capacity for liquids, contains 61·028
cubic inches, or about ·453 of the imperial gallon.

WEIGHT OF WATER.

Maximum density of water 42° Fahrenheit.

Freezing point 32° Fahrenheit, at which point it has expanded $\frac{1}{17}$th of its
original bulk.

62·5 ℔s. avoirdupois = the weight of 1 cubic foot.
  ·03617 .   .   =   .   .   1   „   inch.
  ·434   .   .   =   .   .   1 lineal foot of 1 inch square.
  49·1   .   .   =   .   .   1 cylindrical foot.
  ·02842 .   .   =   .   .   1   „   inch.
  ·341   .   .   =   1 lineal foot 1 inch diameter.
  11·2 imperial gallons =   .   1 cwt.
  224·   .   .   =   .   .   1 ton.
  1·8 cubic feet .   =   .   .   1 cwt.
  35·84   .   .   =   .   1 ton.

1 cubic foot of water = $6\frac{1}{4}$ imperial gallons, and
1 cylindrical foot = about 5.

---

1 circular inch .   .   .   .   =   1·273 square inch.
1 cubic foot   .   .   .   .   = 2200· cylindrical inches.
1   do.   .   .   .   .   = 3300· spherical   do.
1   do.   .   .   .   .   = 6660· conical   do.

METALS

IN CONNEXION WITH THE ARTS, CIVILISATION, AND SOCIAL
PROGRESS.

Gold, the most ancient, the most ductile, and the most valuable of all metals, has always ministered in a large degree to the
perfection of ornamental art. Its opacity, ductility, and durable
brilliancy give it a value in Gold Leaf far beyond the market
price of the metal. The consumption of Gold Leaf is enormous,
the manufacture of which is one of the staple trades of London

and Birmingham. It is not used in medicine, but is valuable and indeed indispensable in the wonderful art of photography. It is difficult to oxidize Gold, it being proof against the effects of nitric acid. Gold is rarely used in the arts in a pure state ; the embellishment of china with Gold is effected by laying the pattern in reduced liquid Gold on to the china : after being burnt in, the brilliancy of the colour is brought out by burnishing the Gold with bloodstones. Gold passes as money in all civilised countries, the standard value being regulated for all the world by the Bank of England.

Silver is readily taken as money in all parts of the world, being the standard in China and some other eastern countries. Its ductility and opacity give it value for embellishing japannery as leaf Silver, for which it is largely used. Large quantities of this metal are likewise consumed by our chemists in making lunar caustic or nitrate of Silver. It is also dissolved very extensively in Sheffield and Birmingham for electro-plating spoons and forks and other articles made in these towns.

Copper goes into consumption for a thousand purposes for which it is more durable and useful than iron, being less liable to oxidize than the former metal. It is the basis of blue vitriol and verdigris, and enters largely into the constitution of brass, this metal being principally composed of spelter with various combinations of Copper. It was formerly used exclusively to protect ships' bottoms; Muntzs's Patent Metal has, however, for years partially superseded it : Copper, however, enters largely into the combination of Muntzs's Rival Sheeting.

Tin is one of the most useful metals we have in manufactures ; it willingly covers and tins over the surface of Iron, effectually preventing oxidization of the former metal, and by this combination Tin Plates are made so useful for the manufacture of culinary utensils and thousands of articles too numerous to mention. The beautiful dish covers made at Wolverhampton, so bright and elegant, are a striking proof of the value of Tin as an elegant covering of Iron. The greatest consumption of Tin takes place in the manufacture of these Tin Plates. Tin is not much used as a medicine, but often consumed as a mixture in the crucible with Gold, Silver, Lead, Copper, and Antimony.

Antimony produces some of our most valuable medicines, and is largely used by the type founders; being one of the most brittle metals it imparts a resisting power to the type, which renders this metal indispensable at the type foundry.

Zinc is the Spelter of commerce, and is the chief metal used in making brass ; it is likewise valuable in the galvanic battery, and, rolled into sheets, is largely used in metallurgical manufactures. A preparation of Zinc is also used for white paint.

Lead is one of the heaviest and most useful metals, and has been known and used from the most ancient times. Being soft and pliable, and melted at a low temperature, it is used largely in the arts, and is an indispensable mixture for most other metals.

Mercury is consumed very largely indeed by the chemists in the manufacture of various mercurial salts and sublimates, pills and ointment. It is useful for many purposes in the arts. The world's produce comes to England ; the consumption has increased so much during the last few years as to enhance the price of Mercury in the market very considerably.

Cobalt, as indigo blue, is the only fast colour for cottons, woollens, and silken goods. Just so Cobalt stands with regard to the blue colour which has to endure the fiery ordeal of the potter's oven ; all the blue colour which we see on china and earthenware is made from Cobalt, hence Cobalt is indispensable in these manufactures.

Iron.—Iron is the hardest,[1] the most abundant, and useful of all metals ; it is deposited in various ways in all parts of the world. All newly discovered countries develop deposits of Iron in the crust of the earth. The best in the world is produced in England, from the estates of the Earl of Dudley, at Dudley ; the noble Earl Granville, at Lillieshall ; and the Duke of Devonshire, at Barrow-in-Furness. The quantities made will be found in the proper place of this book. Its importance in the arts and manufactures is too well known to require comment here ; it forms the basis of some of the most invaluable medicines in the Pharmacopœia, and supplies the material for the great steamships and railroads. In its manufacture, and the industries connected with it, it employs more men, creates a greater amount of wealth, and is of much greater importance to the stability and prosperity of these kingdoms than any other trade, industry, or interest, carried on in the empire.

[1] Of course we include Steel.

IRON is popular with the profession as a tonic in various combinations; perhaps its preparations are more numerously recognised by the Pharmacopœia than any other metal, the most popular preparations are :—

Ferri Carbonas saccharata.[1]
  ,, Citras.
  ,, et Ammoniæ Citras.
  ,, ,, Quiniæ Citras.
  ,, Iodidum.
  ,, Oxidum Magneticum.
  ,, ,, Nigrum.
  ,, Perchloridi Liquor.
  ,, Pernitratis Liquor.
  ,, Peroxidum.
  ,, ,, hydratum.
  ,, Phosphas.
  ,, Potassio tartras.
  ,, Pulvis.
  ,, Sulphas.
  ,, ,, exsiccata.
  ,, ,, granulata.
Ferridcyanide of Potassium.
Ferrocyanide ,, ,,
Ferrum redactum.
  ,, Tartaratum.

[1] We adopt the nomenclature of the British Pharmacopœia, published 1864, in all the names for this and the preparations of other metals

The Tr. Ferri Perchloridi, or Steel Drops, sold in the druggists' shops, is of course made of Iron, and is a most invaluable medicine.

GENERAL TABLE OF METALS.

CONTRACTIONS—Ox. Hy. = melts before Oxy-Hydrogen blow-pipe.　S. F. = melts at highest heat of Smith's forge.

Abbreviations or Symbols	Chemical Equivalents		When and by whom discovered as pure metals	Specific Gravity, or Weight of Water at 60°=1·00	Melting Point Fahrenheit's Thermometer	General Remarks
	Hydrogen =1	Oxygen =100; Hy.=12·5				
1. Gold (Aurum) . Au.	98·33	1229·16	*Known to th Ancients*	19·26	2016°	1. The most ancient metal. Found in small quantities in all parts of the world. At the beginning of this century South America was the largest source of supply; now, the supply comes in the following order—Australia, California, New Zealand, Columbia and other parts of America, and Russia.
2. Silver Argentum Ag.	108·00	1350 00		10·47	1837°	2. Found native in sulphurets, and in most of the lead ores. Our great supply is from South America. Some of our English lead mines are rich in silver.
3. Iron Ferrum . Fe.	28·00	350 00		7·78	2 86°	3. The most universally diffused mineral. Chief supply from Barrow-in-Furness, Millom, Frizington, and other parts of Cumberland, Shropshire, Northamptonshire, Wales, Staffordshire, Yorkshire, and Scotland. Found in various parts of Europe and America; rarely found native. The only metal, except platinum and potassium, capable of being welded, or hammered or pressed together, two several pieces into one solid mass. Used largely for med ca purposes, see page 297.

				Remarks
4. COPPER (*Cuprum*) Cu.	31 66	395·70	8·89	1996°	Found native, frequently in small quantities, and mines wrought in every quarter of the globe. England alone (chiefly Cornwall and Devon) formerly yielded about two-thirds of the entire produce of the world. Copper is used for electro-plating and electro-engraving to a large extent; this metal is likewise the basis of the Cupri Sulphas of the British Pharmacopœia, which is very largely used, and an indispensable salt in every surgery and druggist's shop. It is also consumed largely by the agriculturists. Blue vitriol is a powerful caustic and detergent, and supposed to leave the parts under its influence by contact less liable to inflammation than its more expensive rival Argenti Nitras.
5. MERCURY (*Hydrargyrum*) . . Hg.	100 07	1250·90	13 60	39° below Zero	Found chiefly in Austria and Spain, also in Japan, South America, Mexico, China, &c.; frequently found pure, or sulphuret and other ores 50 to 80 per cent.; rich in pure mercury. Used largely to make mirrors and barometers, at the gold diggings to collect the gold, also in making vermillion, and is the basis of many of the most useful medicines in the British Pharmacopœia, among which may be mentioned Hydrargyri cum cretâ and numerous other salts and preparations, detailed at p. 306.
6. LEAD (*Plumbum*) Pl.	103·56	1294·50	11·35	6 2?	Chief ore, galena or sulphuret of lead. Chief supply, England, Spain, Ireland, Isle of Man, Scotland; sometimes found native in Alston Moor Cumberland. Lead renders great

GENERAL TABLE OF METALS—*Continued.*

Abbreviations or Symbols	Chemical Equivalents		When and by whom used as metals	Specific Gravity, or Weight Water at 60°=1·00	Melting Point, Fahrenheit's Thermometer	GENERAL REMARKS
	Hydrogen =1	Oxygen =100; Hy.=12·5				
7. Tin (*Stannum*). Sn.	58·82	735·24		7·30	442°	7. ...assists ...trace to the ..., the carbonate of lead forms ... white lead of commerce, ...used in oil it is the most durable medium for ...ther ...and paints. In medicine ...it is invaluable. The British Pharmacopœia has the Unguentum plumbi carbonatis and Unguentum plumbi ...the Liquor plumbi ..., and ...for preparations ...well known to the pharmaceutist. ...ill, ...Ma, and Banca are the only sources of supply worth ...ice; found in very limited quantities in ...ther parts of Europe and Asia. Cornish tin, till 837, paid a duty of £4 per ton to the Duke of Cornwall.
8. Antimony (*Stibium*) . . . Sb.	129·03	1612·90	A.D. *Basil Valentine* . . 1490	6·70	810°	8. ...or, *sulphuret.* Found in Hungary, the Hartz Mountains, and in France. Brittle metal; principally used in ...ing type metal, and white ...tal for ...&c. Many valuable medicines are ...red by chemists and pharmaceutists from this ...tal; without it ...ur materia medica would be incomplete. From this we get the Vinum antimoniale, or ...any wine; also Antimonii terchloride liquor, Antimonium sulphuratum,

Antimonium tarta __, Antimonii oxidum, Antimonii oxysulphuretum, Antimonii potassio ... nts, Antimonii sulphuretum ... um, Antimonii ... um præcipitatum, and ...

9. Brittle; of a reddish wh te ... do. ... 8 : ..., lead 5, tin 3, ... fom a mixture that melts in boiling ... Ind in snll ... and combined w th other ors, in various parts of Europe. ... by fom Schneeberg, in Saxony. Used largely in medicine, see p. 306.

10. ... Chief supply, Silesia and ... Since the removal of tho ... duty on foreign, very little is smelted in this ... country, although the ores are found n various ... es. Very malleable at 212° ... hr. Calamine, one of the chief ores of z.nc, was lor g used before it was known as a ... metal. Used largely in medi in, see p. 306

11. Very soft and ... lh of a stee -gray colour; readily oxidises. The deadly poison (used largely in medi in, see p. ... into arsenic of commerce, is the oxide of arsenic; it ... dms any mtl wi ... ith it may be mixed, and is thus used w th lead, 1 to 10, in making shot. Chief supply fom Saxony and Bohemia. The wl ite oxide is destructive to all animal ... The Liquor ... id, nevertheless, is often administered as an effective remedy in ... and p-scribed n very smnl doses son etimes as a ...

9. BISMUTH	Bi.	70·95	886·92	*Agricola* 530	9·80	497°
10. ZINC	Zn.	32·52	406·59	*Paracelsus* 1535	7·00	773°
11. ARSENIC	As.	75·00	93·50	*Brand* 1733	5·88	00°

GENERAL TABLE OF METALS—*Continued.*

Abbreviations or Symbols	Chemical Equivalents		When and by whom discovered as pure metals	Specific Gravity, or Weight Water at 60°=1·00	Melting Point, Fahrenheit's Thermometer	GENERAL REMARKS
	Hydrogen =1	Oxygen =100; Hy.=12·5	A.D.			tonic; some foolish girls are weak enough to take small doses of arsenic with the view of improving their complexion.
12. COBALT . Co.	29·52	368·99	Brandt . .1733	8·53	2800°	12. Reddish gray, and brittle. Little used in the metallic state. Oxide of cobalt, a brilliant blue, much used in the arts.
13. PLATINUM . . Pt.	98·68	123 50	Charles Wood 1741	20·98	Ox. Hy.	13. Very c se, beautiful white metal, most indestructible; nearly as valuable as gold. Very ductile and malleable; less hard than iron. Found in the Ural Mountains, South America, and Spain.
14. NICKEL . Ni.	29·57	369·68	Cronstedt .1751	8·27	2800°	14. From Germany and Sweden. Used in the manufacture of German silver; best, 8 copper, 3 nel, 3½ zinc; com, 8 per, 2 nickel, ¼ zinc. Used also in the manufacture of porcelain.
15. MANGANESE . . Mn.	27·67	345·90	Gahn . . .1774	6 85	S. F.	15. Not used in the metallic state. The manganess of commerce (black oxide) is largely used in the manufacture of bl ng w.
16. TUNGSTEN (Wolfram) . . . W.	94·64	1183·00	Delhuyart . 1781	17·50	Do.	16. White, hard, brittle; found as tungstate of lime, iron, and manganese; not used in the arts.
17. TELLURIUM . . To.	66·14	801·76	Müller . . 1782	6·10	650°	17. Colour, tin white; not used in the arts.
18. MOLYBDENUM . Mo.	47·88	598·52	Hielm . . 1782	7·40	Ox. Hy.	18. White, brittle, and very infusible its native

19. URANIUM . . U.	60·00	750·00	Klaproth . 1789	9·00	B.	su phire was ong aken for an ore of ead, and is frequently found united with lead or. A dk wper, slight metallic lustre; very ; uns with a white light.
20. TITANIUM . . Ti.	24·29	303·66	Gregor . . 1790	5 30	B.	Rare, and very hard; found in slags of iron-smelting , in bright d crystals.
21. M . .. Cr.	28·15	351·82	Vauquelin . 1797	5·90	B.	sh and brittle; not used in the metallic st Its ns form many useful colours to dy, calico-printers, and potters.
22. CLUMBIUM, or TANTALUM . Col.	92·30	1153·72	Hatchett . . 1802	5·60	Burns freely in air	Black wt ; es ustrous under a
23. PALLADIUM . . Pd.	53·27	665·90	Wollaston . 1803	11·80	Do.	nd associated w th pla inum es, ich lit b es It is hard, uti e, and malleable.
24. RHODIUM . . R.	52·11	631·39	Do. . . 1803	10·65	Do.	Whi ish, and wy hard. Has been used for the points of .
25. IR D UM . .. Ir.	98·68	1233·50	Tennant . . 1803	18 68	Do.	Gray and brittle; prod es, in its nis, my brilliant col urs, ehce the amo (ds, the Rainbow).
26. SIUM . . Os.	99·56	1244·49	Do. . . 1803	10 00	Do.	Grayish white, porous, and assuming a m-tallic lustre when compressed.
27. CERIUM . . Ce.	46·00	575·00	Hisinger and Berzelius . 184	...	B.	, brittle; of no use in the s.
28. POTASSIUM (Kalium) . . K.	39·00	487·50	Davy . . 18	0·865	136°	Tl e lightest sold subs n, and a most remarkable metal; has so strong an affinity for oxygen, that it takes fire upon being thrown upon ar. Forn s the metallic base of potash of commerce. This al is in small globular fragmen s, always kept in stoppered bottles red w th spirits of wine to pent spontaneous comb stion; ow ng o its gat fiu y for wa er, a few

GENERAL TABLE OF METALS.—Contin d.

	Abbreviations or Symbols	Chemical Equivalents — Hydrogen =1	Chemical Equivalents — Oxygen =100; Hy.=12·5	When and by whom discovered as pure metals	Specific Gravity or Weight Water at 60°=1·00	Melting Point, Fahrenheit's Thermometer	Remarks.
29. SODIUM (Na rd-nium) . . .	Na.	22·97	287·17	Davy 1807	0·97	190°	...ries ...rn into a basin full of water, ...ase instant ...ustion, giving a brilliant light in a dark ...m.
30. BARIUM .	Ba.	6864	858·01	Do. 1807	4·00	Melts below ...miss	29. The metallic base of ... Soft and malleable, resembling potassium in many features. Thrown on water, it does not ... but is ...ly converted into ... and
31. STRONTIUM . .	Sr.	4384	8·62	Do. 1807	...	...	30. Of a gray colour, but of no use in its metallic state; it forms the base of ... 31. ..., solid, and heavy; rapidly absorbs oxygen; the metallic basis of an earth ... in a pale green mineral at Strontian, in Argyleshire. Nitrate of strontia gives a fine red ... in ...; used to make the red fire in ..., and fireworks at great ...
32.	Ca.	20·00	250·00	Do. 1807	4·0 to 8·0	...	32. A brilliant ... and highly inflammable; the ... base of ... 20 parts ... 8 parts oxygen = common ...
33.	Cd.	55·74	6877	Stromeyer .1818	8·60	442°	33. ..., like tin; too scarce to be used in the arts; ...es the ...ide (yellow) is used as a pigment
34. LITHIUM . . .	Li.	6·43	80·37	Arfwedson .1818	...	...	34. ... obtained by the galvanic ...; ...ly from lithia; an ... ba.

No. & Name	Symbol			Discoverer	Year	Density	Fusibility	Notes
35. SILICIUM · ·	Si.	2·35	266·82	Berzelius	. 1824	1·837	Almost infusible	35. The metallic base of silica, or flint; deep brown powder, [?]te of 1 [?]re.
36. ZIRCONIUM	Zr.	33·62	420·20	Do. · ·	1824	...	...	36. Bla k powder, [?]bling charcoal.
37. ALUMINUM	Al	13·69	171·17	Wöhler · ·	1828	2·58		37. It was very diff[?]ult to procure in the metallic state in any quantity, till [?]ly a French [?]mist has called attention to it, [?]d obtained [?]nt to declare it very malleable; bright [?]d beautiful, like silver, [?]d not affected by the ordinary [?]ids; now used very largely in the arts, particularly at Birmingham.
38. GLUCINUM · ·	Gl.	26·50	331·26	Do · · ·	1828	...	Elt of [?]	38. [?]k gray; the base of glicina, [?]ich is only found in three rare minerals—the emerald, beryl, and euclase.
39. YTTRIUM ·	Y.	32·20	402·61	Do. · · ·	1828	...	...	39.
40. THORIUM ·	Th.	59·50	744·90	Berzelius	. 1829	9·402	Infusible by.	40. Dark iron gray; [?]en ignited in op[?]n air it [?]oes thorina, a fine snow-[?]ite pwder.
41. [MAGNESIUM]	Mg.	12·67	158·35	Bussy	. 1829	2·24	Volatile at white ht.	41. Brilli[?]art [?]d [?]e, the metallic base of magn sia; hard, but ductile, like silver.
42. VANADIUM ·	[Di?]	68·55	856·89	Sefström	. 1830	...	...	42. White, like silver; good [?]ctor.
43. LANTHANIUM ·	Ln.	48·00	600·00	Mosander	. 1838	...	...	43. Dark gray [?]allic powder; not us[?]d in tho arts.
44. DIDYMIUM ·	Di.	48·00	600·00	D). · ·	1811	...	...	44.
45. ERBIUM ·	Er.	...	...	Do. · ·	1843	...	...	45. The metallic base of orbia, a dark yellow erth; [?]l with yttria and turbia.
46. TERBIUM ·	Tr.	...	...	Do. · ·	1843	...	...	46. Hypothetic base of torbia; properties little[?] [?]wn yet.
47. RUTHENIUM ·	Ru.	52·11	652·49	Kalus	. 1844	8·6	...	47. One of the (so- [?]led) noble metals; closely resembles iridium.
48.	Pp.	...	...	H. Rose	. 1845	...	...	48.
49.	Nb.	...	...	Do. · ·	1845	...	...	49. Recently discovered to be identical. *Niobium* is retained as the name.
50.	Il.	Little known yet.	...			...	...	50.
51. NORIUM ·	Nr.	Do.		Svanberg	. 1849	...	...	51.

CHEMICAL.

Preparations and compounds of which Quicksilver forms the chief basis in Pharmacy, all these being recognised by the British Pharmacopœia—

HYDRARGYRUM (*Quicksilver*) :—

Hydrargyri Iodidum Rubrum.
,,　　　,,　　　Viride.
,,　　　Nitrates Liquor Acidus.
,,　　　Oxidum Rubrum.
,,　　　Subchloridum.
,,　　　Ammoniatum.
,,　　　Corrosivum Sublimatum.
Hydrargyrum cum Cretâ.
Pilula Hydrargyri.
Unguentum Hydrargyri.
,,　　　,,　　　Ammoniati.
,,　　　Iodidi Rubri.
,,　　　Nitratis.
,,　　　Oxidi Rubri.

Preparations of which Bismuth forms the basis.— Bismuthum album, synonymous with Bismuthi nitras and Bismuthi subnitras, now prescribed largely by the faculty. Bismuth is often combined with lozenges.

Zinc forms the basis of the following Salts and Preparations:—

Zinci Acetas.	Zinci Oxidum.
,, Carbonas.	,, Suphas.
,, Chloridum.	,, Valerianas.

STRENGTH OF MATERIALS.

COHESIVE STRENGTH.

THE cohesive strength of a body is that force by which its fibres or particles resist separation; therefore, the more particles there are in a body, the greater will be the power requisite to tear them asunder : *hence, the strength of bodies is as the area of their cross sections.*

Note.—The average breaking weight of a Bar of Wrought Iron, 1 inch square, is 25 tons: its elasticity is destroyed, however, by about two-fifths of that weight, or ten tons. It is extended, within the limits of its elasticity, ·000006, or one tenthousandth part of an inch for every ton of strain per square inch of sectional area.

Hence, the greatest constant load should never exceed one-fifth of its breaking weight, or 5 tons for every square inch of sectional area.

Column B, in the following Tables, gives the cohesive strength, in lbs., of the various bodies to which they refer.

General Properties of Metals. Table 1.

Name of Metal.	WEIGHT IN LBS.				STRENGTH.				Expansion in length for 1 degree of heat.
	Of a cube foot.	Of a plate 1 ft. sq. and 1 in. thick.	Of a bar 1 in. sq. and 1 ft. long.	Of a rod 1 in. in diam. and 1 ft. long.	Weight in lbs. required to crush 1 square inch.	Weight in lbs. required to tear under 1 sqe ih.	Value of E.	Value of S.	
					(A.)	(B.)	(E.)	(S.)	
Cast Iron .	450	37·5	3·12	2·45	107,750	17,920	1331	2548	·00000617
Wrought Iron	475	40·5	3·33	2·61	70,000	58,952	1803	2850	·00000698
Steel . .	490	40·8	3·40	2·67	...	130,000	2098	...	·00000636
Copper (cast).	549	45·7	3·81	2·99	116,480	19.072	...	...	·00001430
Gun-metal .	510	42·5	3·54	2·78	...	35,840	714	...	·00001009
Brass (yellow)	523	43·6	3·63	2·85	163,520	17,958	646	890	·00001044
Lead (cast) .	710	59·3	4·94	3·88	7,840	1,824	52	196	·00001593
Zinc (cast) .	439	36·6	3·05	2·40	...	...	990	746	·00001634

General Properties of Timber.   Table 2.

Name of Wood.	Weight in Pounds.		Strength.				Mean diameter of the trunk.	Average length of the trunk.
	Of a cube foot.	Of a bar 1 inch square and 1 foot long.	Weight in lbs. required to crush 1 square inch.	Weight in lbs. required to tear asunder 1 sq. in.	Value of E.	Value of S.		
			(A.)	(B.)	(E.)	(S.)	Ins.	Ft.
Ash	48	·33	8683	14,130	119·	2026	23	38
Beech . . .	44	·30	...	11,500	98·	1556	27	44
Chesnut . . .	55	·38	...	8,100	67·	...	37	44
Elm	35	·24	1284	9,740	50·64	1013	32	44
Fir, Mar Forest .	44	·30	...	6,900	63·	1200	...	...
,, New England .	35	·24	...	10,210	158·5	1102		...
,, Riga . . .	47	·33	...	9,500	90·	1100	20	75
Larch . . .	34	·24	4920	12,240	76·	900	33	45
Mahogany, Honduras	35	·24	...	11,475	115·3	...	} 72	40
,, Spanish .	53	·37	8198	7,560	65·5	...		
Norway Spar . .	36	·25	...	8,320	105·47	1474	15	60
Oak, Adriatic . .	62	·43	...	12,830	70·5	1383	...	...
,, Canadian .	55	·38	...	10,220	155·5	1766	34	53
,, Dantzic . .	47	·33	...	12,720	86·2	1457	...	...
,, English . .	58	·41	9509	11,880	105·	1672	32	42
Pine, pitch . .	41	·29	...	9,800	88·68	1632	...	...
,, red . . .	41	·29	5748	11,840	133·	1341	...	...
Sycamore . . .	38	·26	7082	9,620	75·	...	29	32
Teak . . .	47	·32	...	12,920	174·7	2462	...	...

General Properties of Natural Stones.   Table 3.

Name of the Stone.	Weight in Pounds.		Strength.				Bulk of water absorbed, that of the stone being 1.	Composition.		
	Of a cube foot.	Of a cube yard.	Weight in lbs. on 1 sq. in. producing first fracture.	Weight in lbs. required to crush 1 square inch.	Weight in lbs. required to tear asunder 1 sq. in.	Weight of particles disintegrated in grains.		Silica, per cent.	Carbonate of lime, per cent.	Carbonate of magnesia, per cent.
			(A.)		(B.)					
Sandstone. .	144	3888	3941	5964	772	6·2	·097	96·2	1·1	0·0
Oolite . . .	131	3546	1491	2574	857	8·3	·155	0·4	93·8	2·7
Limestone . .	144	3888	1751	4068	...	10·5	·114	5·0	83·9	4·2
Magnesian Limestone	141	3807	2733	5219	...	1·5	·148	1·7	54·6	40·6
White marble . .	169	4563	4950	6060	551	...	...	1·1	94·5	0·0
Aberdeen Granite .	164	4428	9251	10910	....	......	....	Quartz, Feld-		
Welsh Slate .	172	4644	...	...	11500			spar and Mica.		

TRANSVERSE STRENGTH OF BEAMS, BARS, &c.

If a beam be supported at both ends, and loaded in the middle, it will bend (which is called deflection); and if the load be increased, it will break (which is called fracture).—If a beam two inches deep and one inch broad support a given weight, another beam of the same depth, and double the breadth, will support double the weight; *hence, beams of the same depth are to each other as their breadths* :—again, if a beam two inches deep and one inch broad, support a given weight, another beam of four inches deep and one inch broad, will support four times the weight ;—*hence, beams of equal breadths are to each other as the squares of their depths* :—again, if a beam of a given cross section one foot long, support a known weight, another beam of the same cross section, but two feet long, will support only half the known weight ;—*hence, beams of equal dimensions are to each other inversely as their lengths* ; *therefore, the strength of beams is directly as their breadths and square of their depths, and inversely as their lengths* ; *and if cylindrical, as the cubes of their diameters.*

TRANSVERSE STRENGTH OF TIMBER.[1]

RECTANGULAR BEAMS.

PROBLEM I.

To find the ultimate Transverse Strength of any Rectangular Beam of Timber, fixed at one end and loaded at the other.

RULE.—Multiply the tabular value of S (see Table 2), by the breadth and square of the depth, both in inches, and divide that product by the length also in inches; the quotient will be the weight in ℔s.[2]

Example I.—What weight will it require to break a beam of Fir, the breadth being 2 inches, depth 6 inches, and length 20 feet ?

$$\frac{1100 \times 36 \times 2}{240} = 330 \text{ ℔s.}$$

---

[1] See Barlow's Essay on the Strength and Stress of Timber.—*Art.* 149.

[2] When the beam is loaded uniformly throughout its length, the same Rule will still apply, only the result must be doubled.

Example II.—What is the weight requisite to break a beam of Ash, 7 inches square, 3 feet from the wall?

$$\frac{2026 \times 7^3}{36} = 19,303\tfrac{10}{38} \text{ lbs.}$$

Example III.—What will be the dimensions of a Fir beam 26 feet long, to support a weight of 400 lbs?

$$\frac{312 \times 400}{1100} = 113\cdot5 \text{ the breadth and square of the depth.}$$

Suppose the breadth to be $2\frac{1}{2}$ inches, then $\dfrac{113\cdot5}{2\cdot5} = 51\cdot4$ the

square of the depth, and $\sqrt{51\cdot4} = 7\cdot17$ the depth.

Suppose the depth to be 8 inches, then $8^2 = \dfrac{113\cdot5}{64} = 1\cdot77$ the breadth.

PROBLEM II.

To compute the ultimate Transverse Strength of any Rectangular Beam, when supported at both ends and loaded in the centre.

RULE.—Multiply the tabular value of S by the square of the depth in inches, and four times the breadth; divide that product by the length in inches, and the quotient will be the weight in lbs.

Example I.—What weight will break a beam of English Oak 7 inches broad, 9 inches deep, and 30 feet between the props?

$$\frac{1672 \times 81 \times 28}{360} = 10,534 \text{ lbs.}$$

Example II.—A beam of Beech, 7 inches deep, 4 inches broad, and 10 feet long, supports a weight of 4 tons; what additional weight will require to be added to break the beam?

$$\frac{1556 \times 49 \times 16}{120} = 10,165 - 8960 = 1205 \text{ lbs.}$$

Example III.—What will be the dimensions of a fir beam 30 feet long between the props, to support a weight of 6000 lbs?

$$\frac{6000 \times 360}{1100} = 1963\overset{..}{\cdot}63 \text{ the square of the depth, and 4 times}$$

the breadth.

Supposing the breadth 6 inches,

$$\frac{1963\overset{..}{\cdot}63}{6 \times 4 = 24} - 81\cdot81 \text{ square of the depth, and } \sqrt{81\cdot81}$$

$- 9\cdot5 = $ depth.

Suppose the depth 10 inches,

$$10^2 = 100, \text{and} \frac{1963\cdot63}{100 \times 4} = 4\cdot90 \text{ breadth.}$$

Note 1.—When the beam is uniformly loaded throughout its length, the result must be doubled, *i. e.* it will support double the weight.

Note 2.—When the beam is fixed at both ends and loaded in the middle, one-half of the result must be added; and if the weight is laid uniformly along its length, the result must be tripled.

DEFLECTION OF RECTANGULAR BEAMS OF TIMBER
OR CAST IRON.

To ascertain the amount of Deflection of a uniform Beam, loaded in the middle and supported at both sides.

RULE.—Multiply the cube of the length in feet by the weight in lbs, and divide the product by 32 times the tabular value of E multiplied by the breadth and the cube of the depth, both in inches, and the quotient is the deflection in inches.

Example.—A beam of Ash 10 feet long, 8 inches deep, and 4 inches broad, and loaded with 4000 lbs. in the centre; what is the deflection?

$$10^3 = 1000 \times 4000 = 4,000,000.$$
$$\text{E} = 119 \times 32 \times 4 \times 8^3 = 7,798,784, \text{and}$$
$$4,000,000 \div 7,798,784 = \cdot51 \text{ deflection.}$$

To determine the Dimensions of a Rectangular Beam capable of supporting a required weight, with a given degree of deflection, when fixed at one end.

RULE.—Divide the weight to be supported in lbs. by the tabular value of E multiplied by the breadth and deflection, both in inches; and the cube root of the quotient, multiplied by the length in feet, equals the depth required in inches.

Example.—A beam of Ash is intended to bear a load of 700 ℔s. at its extremity, its length being 5 feet, its breadth 4 inches, and the deflection not to exceed $\frac{1}{2}$ an inch.

Tabular value of $E = 119 \times 4 \times 5 = 238$, the divisor; then $700 \div 238 = \sqrt[3]{2\cdot94} \times 5 = 7\cdot25$ inches, depth of the beam.

To find the Dimensions of a Beam capable of sustaining a given weight, with a given degree of deflection, when supported at both ends.

RULE.—Multiply the weight to be supported in ℔s. by the cube of the length in feet; divide the product by 32 times the tabular value of E multiplied into the given deflection in inches, and the quotient is the breadth multiplied by the cube of the depth in inches.

Note 1.—When the beam is intended to be square, then the fourth root of the quotient is the breadth and depth required,

Note 2.—If the beam is to be cylindrical, multiply the quotient by 1·7, and the fourth root of the product is the diameter.

STATISTICS OF THE SCOTCH IRON TRADE.

Extracted, principally, from the Trade Circular issued by Thomas Thorburn, Esq., Glasgow.

PRODUCTION.

The make of PIG-IRON, *in Great Britain, as nearly as could be arrived at, was as follows :—*

	Furnaces.	Tons.	Annual Average Production per Furnace. Tons.
In 1740 . by 59 .		17,350	294
1760 . . ,, ...		22,000 . .	...
1788 ,, 85		68,000	800
1796 ,, 121		125,000	1033
1806 ,, 169		250,000	1479
1820 . . ,, ...		400,000	
1827 ,, 284		690,000	2429

SCOTLAND ALONE PRODUCED

		Tons.	
In 1827 by 18		36,000	2000
1840 ,, 64		241,000	3765
1845 ,, 88		475,000	5397
1846 ,, 98		570,000	5816
*1847 ,, 100		510,000	5100
1848 ,, 103		580,000	5631
1849 ,, 112		690,000	6160
*1850 ,, 105		595,000	5666
1851 ,, 112		760,000	6785
1852 ,, 113		775,000	6858
1853 ,, 114		710,000	6228
1854 ,, 117		770,000	6581
1855 ,, 121		825,000	6818
1856 ,, 126		832,000	6603
1857 . . ,, 129 .		. 448,000 to June 30th.	

---

* In 1847 a strike took place among the miners, which lasted from July to September, and in 1850 from May to July.

The following Table shows the increase in this branch of the Iron industry in 1854, as compared with 1825:—

DISTRICTS.	No. of Iron Works.		Furnaces in Blast.		Tons of Pig-Iron Produced.	
	1825.	1854.	1825.	1854.	1825.	1854.
Staffordshire . . .	54	72	81	166	171,735	847,600
Shropshire	23	13	36	28	86,320	124,800
Yorkshire	14	14	22	21	35.308	73,444
Derbyshire	9	13	14	25	19,184	127,500
North Wales . . .	9	7	8	9	13,100	32,900
South Wales . . .	37	48	82	121	223,520	750,000
Northumberland, Durham, and North Yorkshire	...	23	...	59	...	275,000
Cumberland and Lancashire.	13	2	2	3	3,000	20,000
Gloucestershire . . .	...	4	...	5	...	21,990
Scotland	9	32	17	118	29,200	796,604
Total	168	228	262	555	581,367	3,069,838

The quantity of Iron. ore required to make a ton of pig-Iron varies according to quality in the different counties : thus, for instance, in Northumberland and Durham, 4 tons is stated to be about the average proportion ; in Yorkshire and Derbyshire, 3 to $3\frac{1}{2}$ tons ; and in Shropshire and Staffordshire, only 3 tons or less.

MR. KENYON BLACKWELL, in a Paper read before the Society of Arts in 1854, on 'The Iron Industry of Great Britain,' gives the following figures, which show the make of Iron in all countries for the previous year :

Countries.	Tons.
Great Britain .	3,000,000
France .	750,000
United States .	750,000
Prussia .	300,000
Austria. .	250,000
Belgium .	200,000
Russia .	200,000
Sweden .	150,000
Various German States .	100,000
Other Countries .	300,000
	6,000 000

By the above table the reader will observe that at nis time England produced as much Iron as all the world besides.

ENGLISH BLAST FURNACES AND ROLLING MILLS NORTH OF THE HUMBER.

From Griffith's 'Iron Trade Circular,' published March 1856. See Griffith's statistics, published in January 1861.

FIRM.	NAME OF WORKS.	FURNACES.			WEEKLY PRODUCE.	
		In.	Out.	Total.	Pigs.	Mallea. Iron.
					Tons.	Tons.
Derwent Iron Company {	Consett .	5	2	7	600	450
	Crookhall .	7		7	840	...
	Bishopwearmouth .	...	...	...	...	400
	Bradley. .	4	...	4	500	...
Bolckow & Vaughan .	Witton Park .	4		4	600	300
Ditto	Middlesbro' .	3	...	3	450	300
Ditto . . .	Eston . .	6	...	6	900	...
Ditto (Elwyn & Co.) .	Eston . .	3	...	3	430	...
Bell Brothers .	Clarence .	3	...	3	400	..
Ditto . . .	Felling .	2	...	2	260	...
Ditto	Wylam .	1	...	1	120	...
Losh, Wilson, & Bell .	Walker . .	4	1	5	560	300
John Carr & Co. .	Tyne Main .	2	...	2	240	...
Weardale Iron Company .	Tow Law .	4	2	6	500	...
Ditto .	Stanhope .	...		1	...	...
Ditto .	Tudhoe .	...	...	...	...	300
James Wakinshaw .	Monkwearmouth .	...	...	...	...	60
Hawks, Crawshay, & Sons	Gateshead .	...		...	...	250
Tyne Iron Company. .	Leamington .	2	...	2	200	40
Hareshaw Iron Company .	Hareshaw .	...	3	3	...	...
Bedlington Iron Company	Bedlington .	...	2	2	...	...
Birtley Iron Company	Birtley .	2	1	3	200	60
Gilkes, Wilson, & Co. .	Middlesbro' .	4	...	4	500	...
Cochrane & Co. . .	Ormsby .		2	4	250	...
B. Samuelson & Co .	Eston . .		...	3	400	...
South Stockton Iron Co. .	Stockton .		...	3	400	...
South Durham Iron Co. .	Darlington .	2	...	2	240	...
Snowdon & Hopkins .	Middlesbro' .	...	...	...	...	200
West Hartlepool Iron Co.	West Hartlepool .	...	3	2	...	...
Total . .		66	17	83	8590	2660

LONDON : PRINTED BY
SPOTTISWOODE AND CO., NEW-STREET SQUARE
AND PARLIAMENT STREET

THE
WELLINGTON COAL & IRON CO.
LIMITED,

MANUFACTURE AT THE

OLD PARK FURNACES,
OLD PARK, SHROPSHIRE,

Best Melting Nos. 1, 2, & 3 Pig Iron,

BEST GREY FORGE
AND
STRONG FORGE PIG IRON,

FROM THEIR

OLD PARK IRON-STONE MINES.

MANAGING DIRECTOR: :
WALTER HOWEL, Esq.
24 Lombard Street, London, E.C.

' GENERAL MANAGER IN SHROPSHIRE
Mr. SAMUEL DANKS,
Malmslee Hall, near Shifnal, Salop.

POSTAL ADDRESS:
The WELLINGTON IRON COMPANY,
Near Shifnal, Salop.

TELEGRAPH STATION
Dawley, Salop.

TESTIMONIALS—*continued.*

'Highfields Foundry and Wrought Iron Works,
'Bilston: November 1850.
'We have for some time purchased Screws and Boxes from Messrs. Colley, which have given us every satisfaction as to their quality and workmanship.
'THOMAS PERRY & SONS.'

'Garndyrriss Iron Works, near Abergavenny,
'June 19, 1850.
'This is to certify we have used Messrs. Colley's Screws and Boxes in our rolling mills for some time, and which have turned out to our satisfaction ; in fact, they are the best we have ever had in our works.
'Per Pro. The BLÆNAVON IRON AND COAL COMPANY.
'THOMAS HEMMING.'

'Messrs. Colley, 'Brierley Hill: July 15, 1858.
'Gentlemen,—We have had one of your 25-ton Lifting Jacks in use for the last nine months, and we have found it to be quite what you represented.
'We are, yours respectfully,
'THE NEW BRITISH IRON COMPANY,
'Per H. M. SIMINCOURT.'

'Park Gate Iron Works, Rotherham :
'April 6, 1858.
'Gentlemen,—We have had in constant use one of your 25-ton Lifting Jacks for the last three years, which has given us great satisfaction. We have never had a Jack at our Works that has done its work so safe as the above, of your make.
'Yours truly,
For SAMUEL BEALE & CO.
'Messrs. Colley.' 'GEORGE S. SANDERSON.

'Freeth Street, Birmingham: Oct. 10, 1867.
'Gentlemen—In reply to your inquiry as to "how the Pins and Boxes in our Roll Frames wear," we have examined them and find that some which have been at work twelve, and others fifteen years, are still in excellent condition, and will probably last yet a further five years. We should state we always work the Pins quite dry. You may use this letter in any way you please.
'Yours truly,
'THOS. WHITFIELD & CO.
'Messrs. Colley & Co.
'West Bromwich.'

'Avon Vale Tin Plate Works.
'Aberavon, Glamorganshire: May 19, 1866.
'Messrs. Colley & Co.
'Dear Sirs,—In reply to yours of the 15th instant, we have let the Screws and Boxes to the Contractors for our Machinery. I shall name your Firm to them and shall have great pleasure in doing so, as your Screws and Boxes gave us satisfaction for many years (in the time of the late Wm. Llavidyn & Sons).
'Yours truly,
'The PORT TALBOT TIN PLATE CO.
'Per THOS. JENKINS.'

TESTIMONIALS—*continued.*

'Bilbao, Spain: May 26, 1868.
'Messrs. Colley & Co.
    'West Bromwich,
    'Dear Sirs,—We are favoured with yours of the 18th instant, and in reply beg
to say that we are not short of any Wrought Iron Pins or Boxes for the present.
Neither can we send you an order for the Lifting Jacks, as we are pretty well supplied
with the two or three we have in use. Should we require any more of these articles
we shall bear you in mind. We are glad to state that the Wrought Iron Pins and
Boxes that you supplied us in June 1866 have given us satisfaction, and remain,
                                    'Dear Sirs,
                                        'Yours truly,
                                            'YBARRA & CO.'

'Eagle Iron Works, West Bromwich :
                                        'July 7, 1849.
    'I have used Messrs. Colley's Screws and Boxes in my rolling mills for some
time, and so far as I have had an opportunity of proving them, I consider them well
manufactured, and they appear to stand their work properly.
                                    'JOHN HARTLAND,
                                        'Per John Hartland, Jun.'

'The Tividale Iron Company, Tividale Iron Works,
                        'Near Tipton: June 10, 1850.
    'Messrs. Colley have supplied us with Pins and Boxes, the quality of which
has turned out to our satisfaction.
                                'THE TIVIDALE IRON CO.
                                    'Per John Hughes.'

'Pontypool.
'Messrs. Colley,
    'Gentlemen,—I am glad to be able to say that the 30-ton Lifting Jack you
sent us does its work well.
                                    'Yours obediently,
                                        'W. WILLIAMS.
'Newport, Monmouthshire :
    'Oct. 31, 1855.'

'Pontymister Works, near Newport,
            'Monmouthshire ; March 7, 1857.
'Messrs. Colley,
    'Hope Works, West Bromwich,
    'Gentlemen,—We have just received the Lifting Jack, which gives satisfaction,
and we enclose our cheque, £15, for which please send receipt. We are not in want
of any Pins and Boxes at present. You did not answer our inquiry about Crab
Winches.
                                    'Messrs. BANKS & CO.'

TESTIMONIALS—*continued.*

'Tividale Iron and Tin Plate Works,
'Near Tipton : April 21, 1858.
'Messrs. Colley,
'Gentlemen,—We have had one of your 20-ton Lifting Jacks in constant use at these Works, and we have no hesitation in saying it is in every respect quite equal to your representations.
'Yours truly,
'Hope Works, West Bromwich.'
'BUDD & CO.

'Pontymister Works, near Newport,
'Monmouthshire : May 3, 1860.
'Messrs. Colley,
'Hope Works, West Bromwich,
'Gentlemen,—We have your favour of 2nd instant, but at present are not in want of Screws and Boxes or Lifting Jacks ; those we have had from you have always given great satisfaction, and when we are in want, will write you.
'Yours respectfully.
'BANKS & CO.'

'From Messrs. Bills & Mills.
'We have been using for some time Boxes and Pins which prove to be very good—they answer our purpose—from Messrs. Colley, West Bromwich.
'P. W. BOOTH.
'Darlaston Green Iron Works :
'July 16, 1849.'

'Crook Hay Colliery and Iron Works,
'West Bromwich : November 1849.
'We have great pleasure in stating that we have used for nine years the Wrought Iron Pins and Boxes manufactured by Messrs. Colley, and have found them to answer our purpose.
'THOMAS DAVIES & SONS.'

'Pontier Caerleon Tin Works : May 17, 1862.
'This is to certify that we have used Messrs. Colley's Pillars and Boxes ,for fifteen years, and during that time we have not broken one; in fact, they are as good now as when we put them up; and we have no hesitation in saying that they are the best we ever had.
'For JOHN JENKINS & CO.
'Edward Francis.'

'Tividale Iron and Tin Plate Works,
'Tipton : Oct. 11, 1867.
'We have for some years used the Pins and Boxes manufactured by Messrs. Colley & Co., and can testify to their very good quality; they are the best we can get to answer our purpose.
'BUDD & CO.
'Per W. H. Jones.'

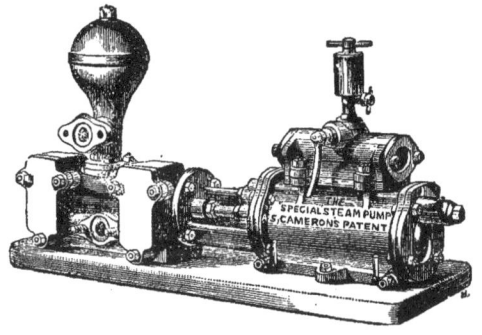

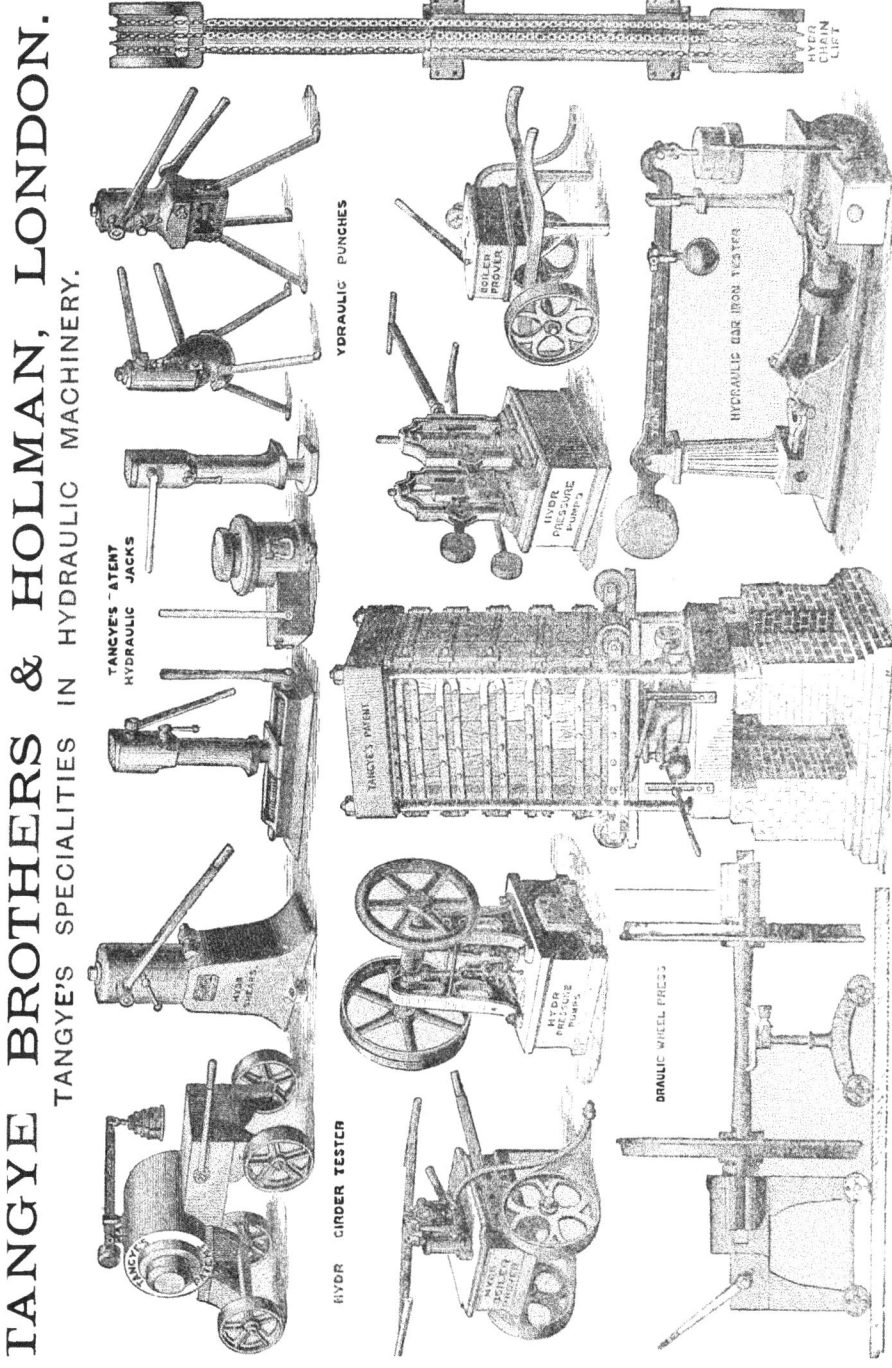

TANGYE BROTHERS & HOLMAN, LONDON.

TANGYE'S SPECIALITIES IN HYDRAULIC MACHINERY.

HYDR CHAIN LIFT

HYDRAULIC PUNCHES

BOILER PROVER

HYDRAULIC BAR IRON TESTER

HYDR PRESSURE PUMPS

TANGYE'S PATENT HYDRAULIC JACKS

TANGYE'S PATENT

HYDR PRESSURE PUMPS

DRAULIC WHEEL PRESS

HYDR GIRDER TESTER

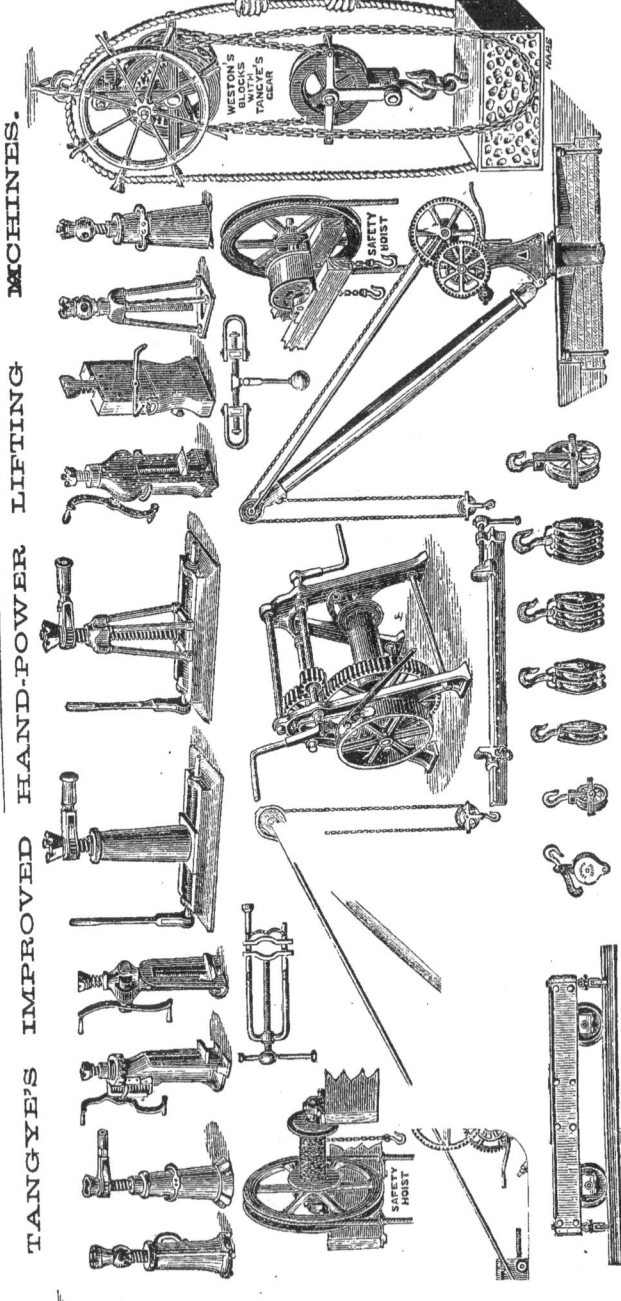

PARIS

UNIVERSAL EXPOSITION,

1867.

THE GOLD MEDAL

WAS AWARDED TO

THE BARROW

HÆMATITE STEEL COMPANY,

LIMITED,

For Excellence in Steel Manufactures.

BARROW

𝕭𝖆𝖒𝖆𝖙𝖎𝖙𝖊 𝕾𝖙𝖊𝖊𝖑 𝕮𝖔𝖒𝖕𝖆𝖓𝖞,

LIMITED.

CAPITAL ONE MILLION.

CHIEF OFFICES AND WORKS:

BARROW IN FURNESS, LANCASHIRE.

LONDON OFFICE:

14 GREAT GEORGE STREET, WESTMINSTER.

DIRECTORS.

HIS GRACE THE DUKE OF DEVONSHIRE (Chairman).
LORD FREDERICK C. CAVENDISH, M.P.
HENRY WILLIAM SCHNEIDER, Esq.
WILLIAM CURREY, Esq.
FREDERICK ILTID NICHOLL, Esq.
SIR JAMES RAMSDEN (Managing Director).
JOHN FELL, Esq.

SECRETARY.

H. L. JONAS, Esq.

MANAGER.

J. T. SMITH, Esq.

Albion Works,

WILLENHALL (Wolverhampton):

JULY 3, 1872.

BY ROYAL LETTERS PATENT

GREAT BRITAIN, FRANCE, AND BELGIUM.

DEAR SIRS,

We take the liberty of again calling your attention to our newly-discovered Oxide, for the annealing of Malleable Iron Castings (Tildesley's Patent Compound). Since forwarding you our Circular of February 6th, it has been very satisfactorily adopted both at home and abroad. At home, while some have only in part adopted its use (mixing it with the Red ore), others, both in Staffordshire, Yorkshire, Lancashire, and Monmouthshire, have entirely abandoned the use of Hæmatite ore, and they assure us that our Oxide is far preferable. Abroad, we have a depôt both in France and Belgium, and we are well satisfied with the sales we have made in those Countries.

When you take into account the present price of Hæmatite ores you will at once see that ours, apart from its superior strength, has the advantage.

Any explanation we can make, or instructions we can give, will be rendered with pleasure.

We may here state that we have ourselves Seven Annealing Ovens in constant use, and now for nearly a year our Oxide alone has been used.

Soliciting your commands, direct or through our Agents,

We are, Sirs,

Your obedient Servants,

JOHN HARPER & CO.

To be obtained only for Annealing purposes at these Works, or through our Agents:—

Messrs. THOMAS COX & CO. ....BIRMINGHAM.

Messrs. G. & W. UNDERHILL ........................WOLVERHAMPTON.

Messrs. CARRICK & BRICKBANK......Clarence Street, MANCHESTER.

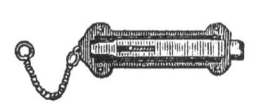  

MALLEABLE IRON CASTINGS.

JOHN HARPER & CO.

IRONFOUNDERS, ETC.

ALBION WORKS,

WILLENHALL, STAFFORDSHIRE;

54½ BISHOPSGATE STREET WITHIN,

LONDON, E.C.

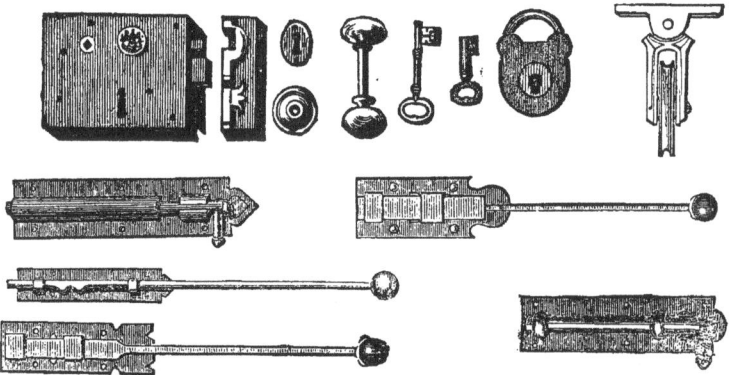

MANUFACTURERS OF

LOCKS, LATCHES, AND DOOR BOLTS:

Axle Pulleys, Door Knockers, Galvanized Signal Pulleys.

ALL KINDS OF

STAMPED, PRESSED, AND FORGED IRON WORK.

ENGINEERING WORKS, POOLE, DORSET

S. LEWIN'S REVISED PRICE LIST.

SUBJECT TO ALTERATION WITHOUT NOTICE.

Portable Engines.

Pages in Catalogue, 4 and 5.

Horse Power.	Number of Cylinders.	Diameter of Cylinders.	Stroke.	Price.
4	1	$6\frac{3}{4}$ inches.	10 inches.	£198
5	,,	$7\frac{5}{8}$,,	12 ,,	220
6		$8\frac{3}{8}$,,	,,	242
7		9 ,,	,,	258
8	,,	$9\frac{5}{8}$,,	,,	280
10	,,	$10\frac{7}{8}$,,	14	324
8	2	each $6\frac{3}{4}$,,	12	320
10	,,	,, $7\frac{5}{8}$,,	,,	352
12		,, $8\frac{3}{8}$,,	,,	407
14		,, 9 ,,	14	456
16		,, $9\frac{5}{8}$,,	,,	500
18		,, $10\frac{1}{4}$,,	16	560
20		,, $10\frac{7}{8}$,,	,,	599
25		,, 12 ,,		726
30	..	,, 13 ..	.. ..	858

Stationary Steam Engines, on Multitubular Boilers.

Page in Catalogue, 7.

SINGLE CYLINDER.

Horse Power.				Price.
4	...	...	...	£193
5	...	...	...	210
6	...	...	...	232
7	...	...	...	248
8	...	...	...	270
10	...	...	...	309

DOUBLE CYLINDER.

Horse Power.				Price.
8	...	...	...	£305
10	...	...	...	335
12	...	...	...	390
14	...	...	...	440
16	...	...	...	485
18	...	...	...	544
20	...	...	...	583
25	...	...	...	704
30	...	...	...	830

Steam Launch Engines.

Prices on Application. *Page in Catalogue,* 3.

Traction Engines.

SINGLE CYLINDER.

Horse Power.				Price.
6	...	...	...	£405
8	...	...	...	484
10	...	...	...	555

DOUBLE CYLINDER.

Horse Power.			Price.
10	...	...	£583
12	...	...	660
14	...	...	769

Tramway Locomotive Engines.

Prices and Particulars on Application.

Lewin's Prize Portable Engine.

Prices of these Engines are under :—

SINGLE CYLINDER.		DOUBLE CYLINDER.	
4-Horse Power... ... ... £176		8-Horse Power...	
5 ,, ,, ... 195		10 ,, ,,	£316
6 ,, ... 216		12 ,, ,,	364
7 ,, ... 230		14 ,, ,, . 400	
8 ,, ... 250		16 ,, ,, ⌐.. . 450	
10 ,, ... 290		18 ,, ,, . 500	
		20 ,,	520
		25 ,,	620
		30 ,,	775

c

Lewin's Vertical Engine Fixed on Boiler.

S. LEWIN'S REVISED PRICE LIST—*continued.*

Horizontal Fixed Engines.

Pages in Catalogue 8 and 9.

CLASS I.			CLASS II.		
Horse Power.	Engine only.	Engine and Cornish Boiler.	Horse Power.	Engine only.	Engine and Cornish Boiler.
4-H.P.	£82	£165	4-H.P.	£68	£150
6 ,,	110	220	6 ,,	80	190
8 ,,	137	275	8 ,,	110	247
10 ,,	165	330	10 ,,	130	295
12 ,,	190	380	12 ,,	160	350
14 ,,	215	430	14 ,,	190	405
16 ,,	242	484	16 ,,	210	452
18 ,,	275	550	18 ,,	220	495
20 ,,	300	600	20 ,,	235	535
25 ,,	360	720	25 ,,	290	630
30 ,,	423	847	30 ,,	325	748
35 ,,	484	968	35 ,,	350	800
40 ,,	550	1100	40 ,,	390	870

Estimates given for Larger Engines.

If Condensers are Fitted to the above.

Up to 10-Horse Power ... ... £42   Up to 25 to 30-Horse Power £95
,, 12 & 14-Horse Power   58   ,, 40 to 50-Horse Power 130
,, 16 & 20   ,,   75

Page in Catalogue 10.

ESTIMATES GIVEN FOR CORNISH, CATER, OR OTHER BOILERS.

Vertical Engines.

Pages in Catalogue 11 and 12.

ON BASE PLATES.	Price.	ON WROUGHT-IRON TRUCKS WITH FOUR WHEELS.	Price.
Horse Power.		Horse Power.	
1½	£82	2¼	£121
2½	110	3	134
3	121	4	159
4	143	5	184
5	170	6	209
6	192		
7	220		
8	236		
9	258		
10	275		
12	319		

S. LEWINS REVISED PRICE LIST—*continued.*
Thrashing Machines and Portable Engines.
Pages in Catologue 13, 14, and 15.

H.P. of Engines required	SINGLE BLAST MACHINE		TOTAL	H.P. of Engines required	FINISHING MACHINE		TOTAL
		£				£	
4	Engine..................... 198 4'6" Small Size 115		£313	4	Engine................... 198 3'6" Machine 125		£323
5	Engine.................... 220 4'6" Small size 125		£345	5	Engine................... 220 4'0" Small size 135		£355
6	Engine.................... 242 4'6" Large size 135		£377	6	Engine .... 242 4'6" Small size 145		£387
7	Engine.................... 258 4'6" Large size 135		£393	7	Engine................... 258 4'6" Large size 160		£418
8	Engine.................... 280 5'0" Machine 145		£425	8	Engine................... 280 4'0" Large size 160		£440
10	Engine.................... 324 5'0" Machine 145		£469	10	Engine................... 324 5'0" Machine 170		£495

Patent Thrashing Machines Combined with Straw Elevator, to deliver at any angle.
Prices on Application.

Horse-Power Thrashing Machines.
Page in Catalogue, 15.

Horse Power.				Width of Drum.					Price.
2-H.P.	...	...	...	24 in.	...	...	...	...	£50
3 ,,				30 in.	...	...	...	...	70
4 ,,	...		...	42 in.	...	...	...	...	80
4 ,,	...	...	...	48 in.		...	...	...	87

Each Price includes Machine and Horse-Gear mounted on 4-wheel Carriage.

These Machines can be fitted with Pulleys to be driven by Steam Power if required.

Patent Stacker and Elevator.
Pages in Catalogue, 16 to 21.

Price Complete	...	...	...	...	...	...	...	...	...	£50
Horse-Gear extra	...	...	...	...	...	...	...	...	...	8

Grinding Mills.
Pages in Catalogue, 22 and 23.

	PRICE OF EACH MILL.			
	2 ft. 8 in. Diameter.	3 ft. Diameter.	3 ft. 6 in. Diameter.	4 ft. Diameter.
If fitted with Derbyshire Grey Stone ...	£53	£60	£72	£88
,, French Bedstone and Grey Runner	58	67	82	99
,, French Stones	64	75	88	114
If provided with Loose Pulley			£3 10s. extra.	
,, Improved Crane to lift the Running Stone for Dressing £7 10s. ,,				

Price for Portable Mills on application.

Cement Mills.
Prices and Particulars on Application.

Mortar Mills.
Page in Catalogue, 29.

5 ft. pan ...	...	...	...	...	...	...	...	...	...	£90
6 ,, ,, ...	...	...	...	...	...	...	...	...	...	100
7 ,, ..	...	...	...	...	...	...	...	...	...	120
8 ,, ...	...	...	...	...	...	...	...	...	...	140
9 ,, ...	...	...	...	...	...	...	...	...	...	160

Clay and Rough Stuff Runners.
Prices on Application. *Page in Catalogue,* 28.

Patent Pipe, Brick, and Tile Machines.
Pages in Catalogue, 26 *and* 27.

Large Size ...	...	...	...	...	...	...	...	...	...	£175
Small Size ...	...	...	...	...	...	...	...	...	...	135

Draining Pipe, and Tile Machine.
Page in Catalogue, 28.
Price with One Die   ...   .   £50.

Prices for Brick Machinery on Application.
Page in Catalogue, 30.

Self-Acting Saw Bench.
Page in Catalogue, 24.

Size of Table, 7 ft. long, by 3 ft. 3 in., 48 in. Saw ...	...	...	...	£75	0*s.*
Movable Carriage, 20 ft. of Rails for same ...	...	...	...	...	25	0*s.*
Boring Table, fitted with Rising Apparatus	...	...	...	3	0*s.*
Spindle bored up, and fitted with Set of Bits and Augers	...	...	...	1	10*s.*

Saw Bench.
Page in Catalogue, 25.

Table 4 ft. by 2 ft.. with 24 in. saw ...	...	...	...	£18	0*s.*
,, 5 ft. by 2½ ft. ,, 30 in. ,,	...	...	...	...	30	0*s.*
Boring Table, fitted with Rising Apparatus, extra ...	...	...	...	2	15*s.*
Spindle bored up and fitted with Set of Bits or Augers	...	...	...	1	5*s.*

Portable Steam or Hand Cranes ...	...	...	...	..	
Patent Hay and Waggon Loader ...	...	...	...	..	Prices and Particulars
Patent 2 and 3 Furrow Ploughs ...	...	...	...	..	on Application.
Stone or Ore Crushing Machine ...	...	...	...	...	

ESTIMATES GIVEN FOR WINDING, PUMPING, AND OTHER GEAR.

A rise of 10 per cent. on all Goods not named in this List.

LILLESHALL COMPANY,

COAL AND IRON MASTERS'

ENGINEERS, ETC.

SHIFNAL, SHROPSHIRE.

MANUFACTURERS

OF ALL DESCRIPTIONS OF

MILL, FORGE AND MINING MACHINERY,

AND

STEAM HAMMERS.

ISAAC JENKS AND SONS

MINERVA & BEAVER IRON, STEEL, & SPRING WORKS,

WOLVERHAMPTON.

MANUFACTURERS OF CAST, SPRING, BLISTER, AND OTHER STEEL

RAILWAY SPRINGS, MERCHANT BARS AND SHEETS, WIRE RODS, ETC., ETC.

TAPER BRAKE LEVER BARS A SPECIALITY.

BRAND OF STEEL JENKS. BRAND OF IRON, 'BEAVER.'

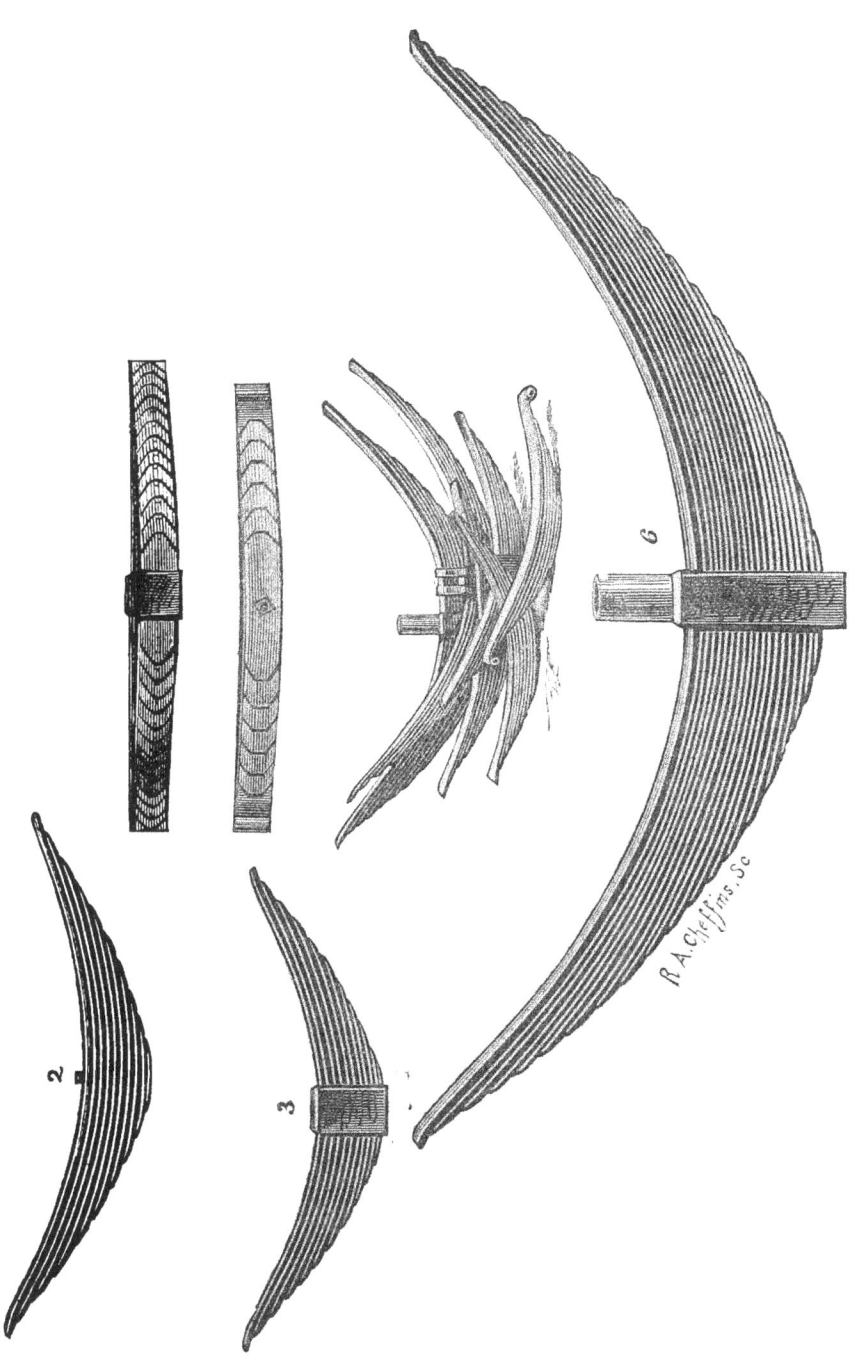

R.A.Cheffins.Sc

MOORE AND MANBY,

DUDLEY.

LONDON OFFICE:

3 BILLITER SQUARE, FENCHURCH STREET, E.C.

CONTRACTORS TO HER MAJESTY'S GOVERNMENT

AND

THE COUNCIL OF STATE

FOR

INDIA.

REGISTERED TRADE MARKS.

M. & M.   **SAXON.**

DESCRIPTION OF MANUFACTURED IRON
OF BEST QUALITIES.

Flat Bars from $\frac{3}{8}$ to 12 inches wide.

Round ditto, from $\frac{1}{8}$ to 8 inches diameter.   Square ditto, from $\frac{1}{8}$ to 5 inches.

Half-round, Feather and Square Edge, to 6 inches wide.

Beveled, Octagon, Hexagon, Oval, Moulding, and every other description of Fancy Iron.

Best, Best Best, and Treble Best Rivet Iron, Plating Bars, &c.

Hoop and Strip Iron from $\frac{1}{2}$ to 10 inches wide.

Sheets—Single, Double, and Lattin.

Roofing Sheets—Corrugated and Galvanized Iron.

Nail Sheets and Hoops, Nail Rods, and Flat Slit Rods.

Boiler Plates—Best, Best Best, and Treble Best; all sizes.

Gasometer and Tank Plates; all sizes.

Ship, Bridge, Girder, and Flitch Plates; all sizes.

Ribbed and Chequered Foot Plates; all sizes.

Canada and Tin Plates, Coke and Charcoal Sheets, &c.

Angle, Equal, and Unequal Sided, and Double Angle.

Tee, Equal, and Unequal Sided, and Double Tee.

Sash Bars and Trough Iron of various sections.

Rolled Girder, Joist, and Beam Iron; all sizes.

Bulb, Bulb Angle, Bulb Tee, and Deck Beam Iron.

Fencing and Telegraph Wire, Black and Galvanized.

Contractors' Permanent, Bridge, and Tram Rails.

Locomotive, Coach, Carrriage, and Waggon Tyres.

Locomotive and other Fire Bars of various sections.

Railway Axles, Forgings, Use Iron, and Stores of all descriptions.

Railway Spikes, Fish Plates, Bolts and Nuts.   Ironwork of all kinds.

Rivets for Shipbuilding and best Boiler Work.

Best Yorkshire Iron supplied of the various brands.

Hot and Cold Blast Melting and Forge Pig Iron.

Rolls turned for irregular sizes of Iron according to agreement.

All information as to Prices &c. can be obtained at

3 BILLITER SQUARE, LONDON,
OR AT DUDLEY.

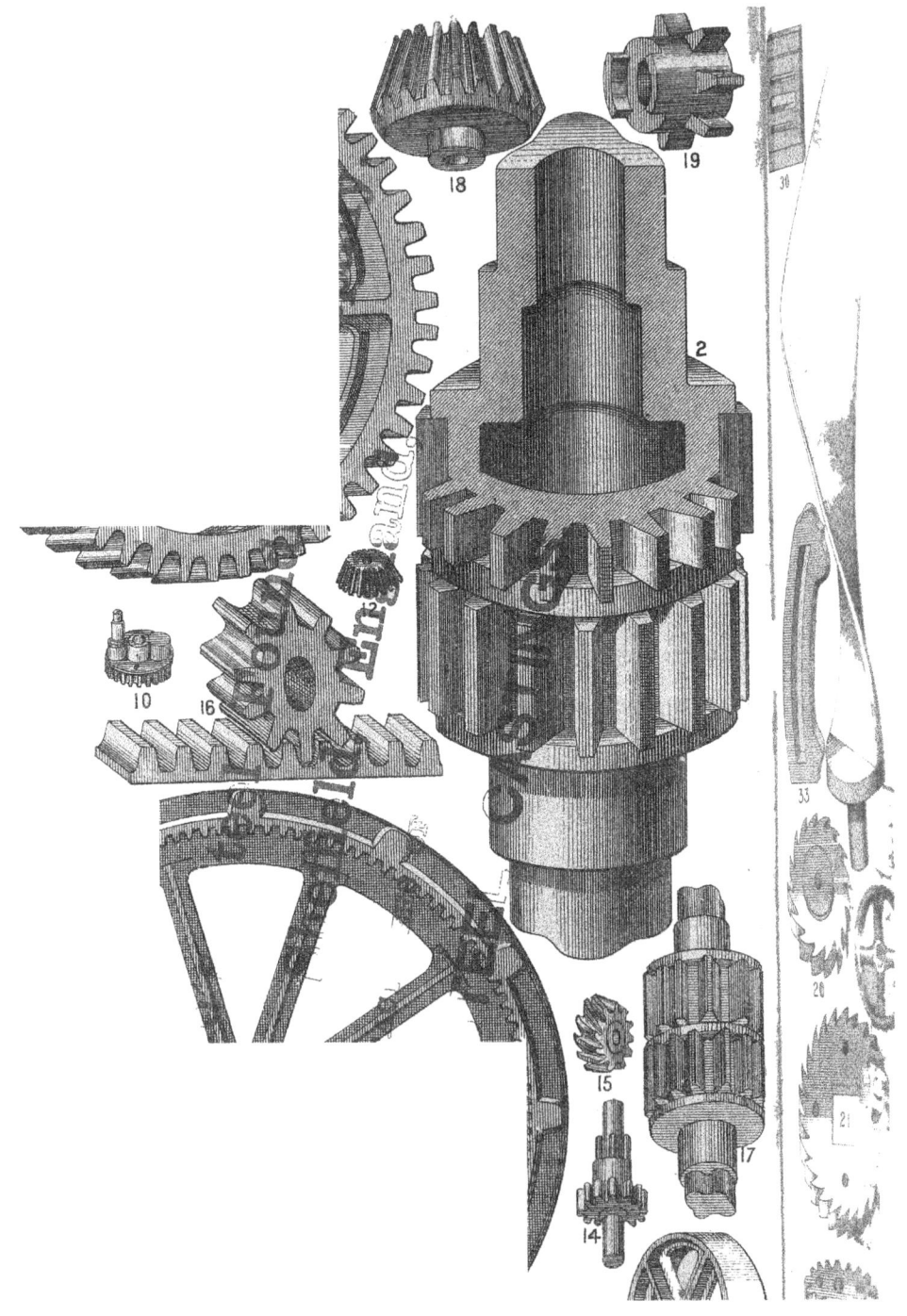

18

19

30

2

10   16

33

15

17

14

28

21

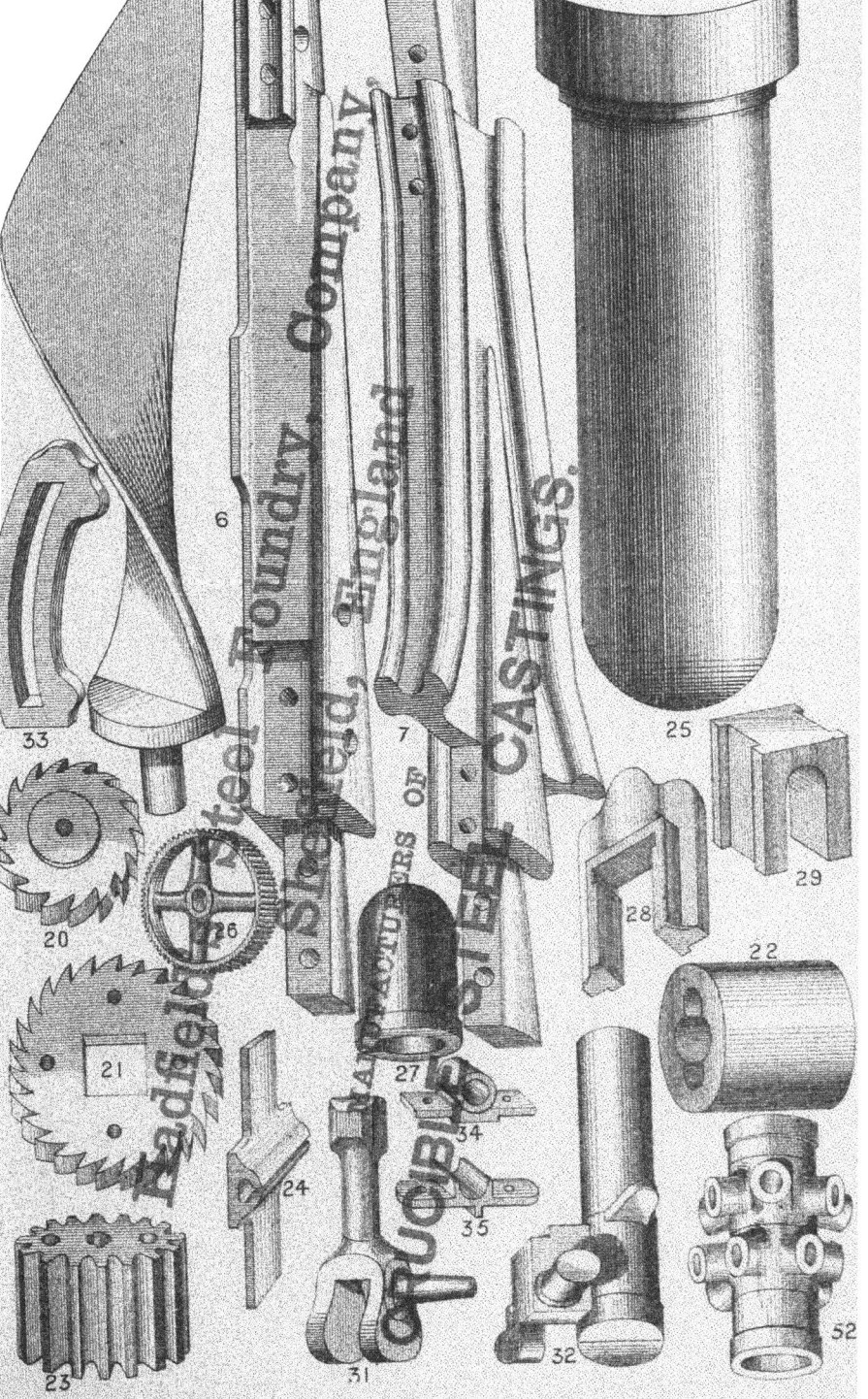

Hadfields Steel Foundry Company

Sheffield, England

MANUFACTURERS OF CRUCIBLE STEEL CASTINGS.

33

6

7

25

29

28

22

20

26

21

27

34

24

35

32

52

23

31

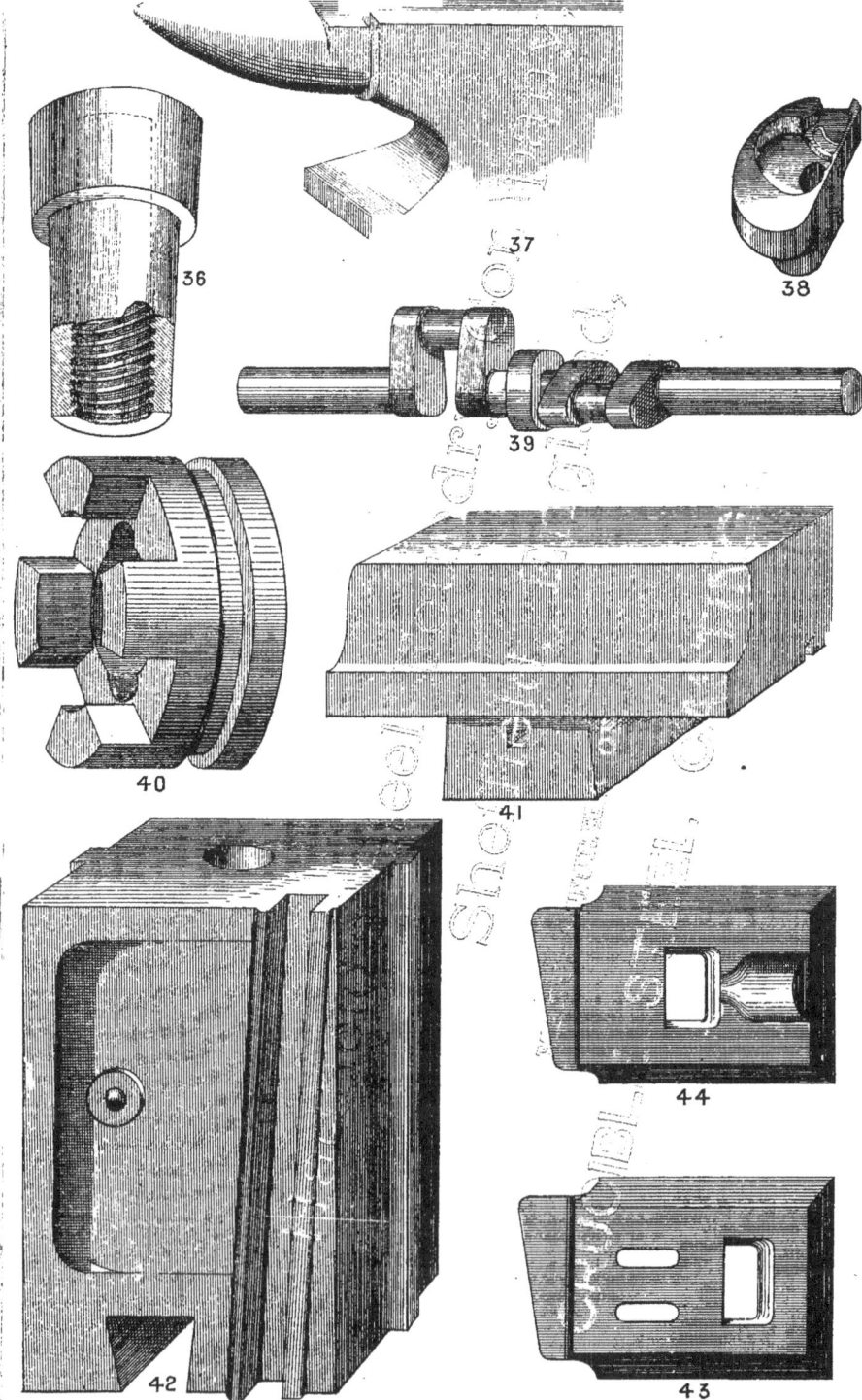

36

37

38

39

40

41

42

43

44

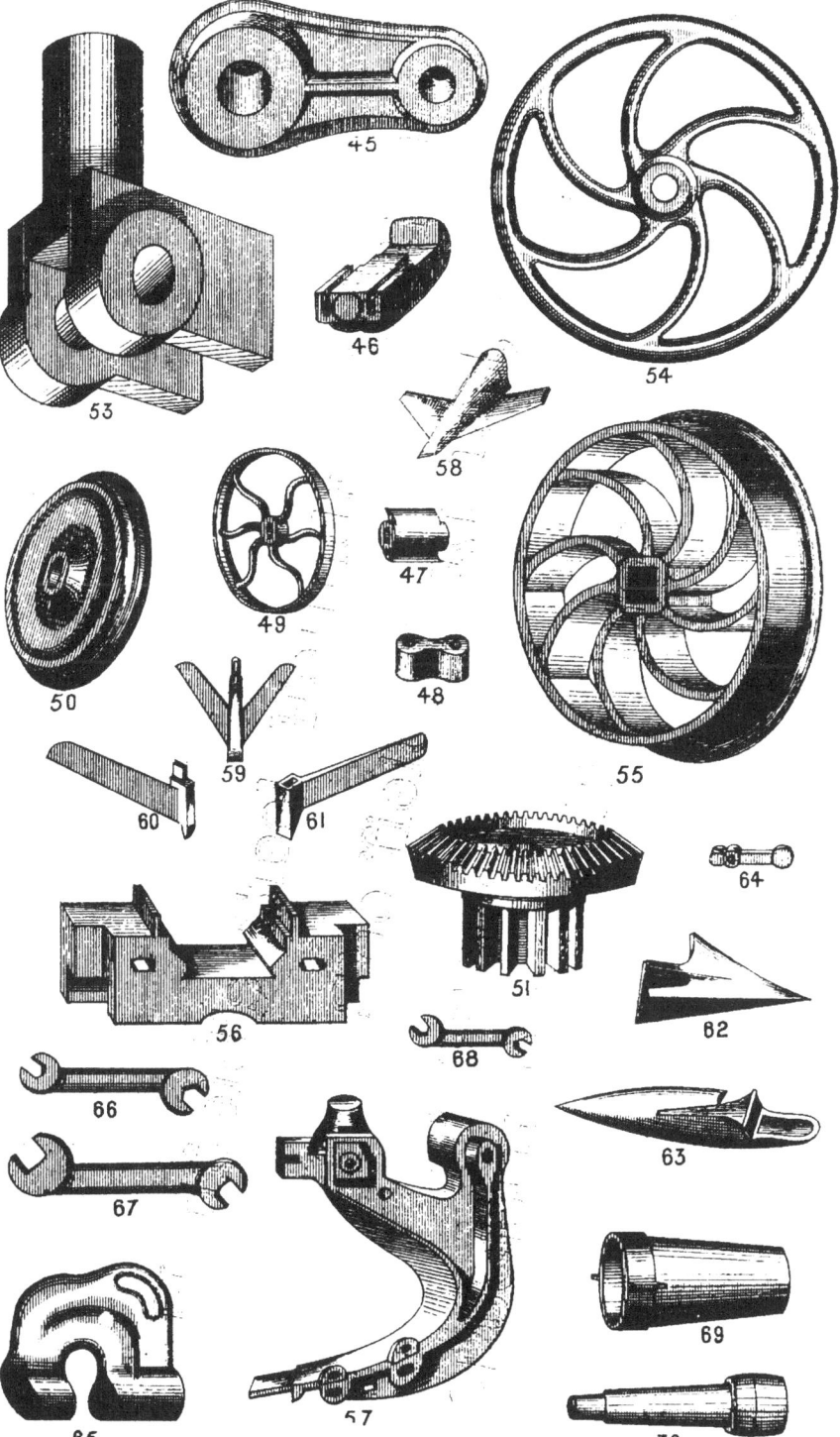

45

53

46

54

58

49

47

50

59

48

55

60

61

51

64

56

62

68

66

63

67

57

85

69

70

THE above Engraving represents Mr. R. HADFIELD's PATENT IMPROVED DOUBLE-DISC RAILWAY WHEEL, the Tyre of which is Steel or Iron, with a Metal Centre *welded* thereto. These Wheels have created considerable attention during the last nine months among Railway Engineers and others.

EDWARD DAVIES,

CROWN'

GALVANIZED IRON WORKS,

WOLVERHAMPTON.

ESTABLISHED 1838.

EDWARD DAVIES—*Continued.*

MANUFACTURER

OF

GALVANIZED CORRUGATED SHEETS

FOR ROOFING PURPOSES,

IN ALL SIZES OF CORRUGATION.

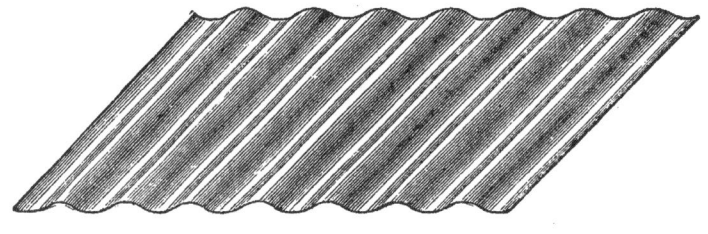

THE <u>ONLY PROPRIETOR</u> AND MANUFACTURER OF THE
REGISTERED

'CROWN BRAND'

PLAIN FLAT SHEETS, MADE EXPRESSLY FOR WORKING UP

Of 'BEST,' 'BEST BEST,' and CHARCOAL QUALITIES.

IMPROVED

GALVANIZED ROOFING TILES,

EDWARD DAVIES—*Continued.*

TINNED SHEET IRON,

GALVANIZED IRON RIDGING,

GALVANIZED HOOPS.

Manufacturer of all Descriptions of Galvanized Iron Goods.

GALVANIZED FITTINGS

FOR

SHEETS, TILES, &c.

GALVANIZED TINNED MACHINE-MADE GUTTERS,

RAIN-WATER PIPES, &c.

GALVANIZED IRON ROOFS,

AND

BUILDINGS FOR HOME & EXPORT

Fixed complete, or fitted for Erection.

WOLVERHAMPTON—CROWN WORKS.

LIVERPOOL— 17 SWEETING STREET.

LONDON—ST. CLEMENT'S HOUSE.

D

PARTICULARS & SIZES OF IRON

MANUFACTURED BY

ROBERT HEATH & SON

BIDDULPH VALLEY,

NORTON AND RAVENSDALE

IRON WORKS,

NORTH STAFFORDSHIRE.

BRANDS.

R. H. ✠ & BIDDULPH. | R. D. ✠ & RAVENSDALE.

PLATES.

All qualities not thinner than $\frac{1}{8}$ of an inch, or thicker than $1\frac{1}{8}$ inches, or wider than $5\frac{1}{2}$ feet.

ROUNDS.

From $\frac{3}{16}$ of an inch to 6 inches.
Sizes vary by $\frac{1}{32}$ up to $1\frac{1}{8}$ inches.
,,   ,, $\frac{1}{16}$ from $1\frac{1}{8}$ to $2\frac{1}{2}$,,
,,   ,, $\frac{1}{8}$,, $2\frac{1}{2}$,, 3   ,,
,,   ,, $\frac{1}{4}$,,   3   ,, 6   ,,

SQUARES.

From $\frac{3}{16}$ of an inch to 6 inches.
Sizes vary by $\frac{1}{32}$ up to $1\frac{1}{8}$ inches.
,,   ,, $\frac{1}{16}$ from $1\frac{1}{8}$ to 2   ,,
,,   ,, $\frac{1}{8}$,,   2   ,, 3   ,,
,,   ,, $\frac{1}{4}$,,   3   ,, 6   ,,

FLATS.

From $\frac{3}{8}$ of an inch to 8 inches wide.
The sizes vary in width by
$\frac{1}{8}$ from $\frac{3}{8}$ to 2 inches, thickness not under $\frac{3}{16}$
$\frac{1}{4}$,, 2 ,, 5   ,,   ,,   ,,
$\frac{1}{2}$,, 5 ,, 8   ,,   thickness not under $\frac{1}{4}$

HOOPS AND STRIPS.

From $\frac{1}{3}$ inch to $1\frac{1}{8}$ inches, not thinner than 23 W.G.
,, $1\frac{1}{4}$,, ,, 2   ,,   ,,   20   ,,
,, $2\frac{1}{8}$,, ,, 3   ,,   ,,   18   ,,
,, $3\frac{1}{8}$,, ,, 5   ,,   ,,   15   ,,
Waved $1\frac{1}{8}$ inches, not thinner than 18 W.G.

OCTAGON IRON.

$\frac{15}{32}$ to $\frac{5}{8}$ inches.
The sizes vary by $\frac{1}{32}$.

CAN TOP.

$\frac{1}{2}$, $\frac{5}{8}$ and $\frac{3}{4}$ inch.

TEE IRON.

Size.			Thickness.			Size.			Thickness.		
8	×	$4\frac{5}{8}$	$\frac{5}{8}$	to	$\frac{3}{4}$	4	×	2[illegible]	$\frac{3}{8}$	,,	$\frac{1}{2}$
6	×	4	$\frac{7}{16}$	,,	$\frac{5}{8}$	4	×	2	$\frac{5}{16}$	,,	$\frac{7}{16}$
6	×	$3\frac{5}{8}$	$\frac{1}{2}$	,,	$\frac{5}{8}$	$3\frac{1}{2}$	×	$3\frac{1}{2}$	$\frac{5}{16}$	,,	$\frac{1}{2}$
6	×	$3\frac{1}{2}$	$\frac{7}{16}$	,,	$\frac{5}{8}$	$\frac{1}{2}$	×		$\frac{3}{8}$	,,	$\frac{1}{2}$
6	×	3	$\frac{3}{8}$	,,	$\frac{5}{8}$		×		$\frac{3}{8}$	,,	$\frac{1}{2}$
6	×	$2\frac{1}{2}$	$\frac{3}{8}$	,,	$\frac{5}{8}$		×	$\frac{1}{2}$	$\frac{3}{8}$	,,	$\frac{1}{2}$
5	×	5	$\frac{1}{2}$	and	$\frac{5}{8}$		×		$\frac{5}{16}$	,,	$\frac{1}{2}$
5	×	$4\frac{1}{2}$	$\frac{1}{2}$	,,	$\frac{5}{8}$	$\frac{3}{4}$	×	$\frac{3}{4}$	$\frac{5}{16}$		1
5	×	4	$\frac{1}{2}$	,,	$\frac{5}{8}$	$2\frac{3}{4}$	×	$2\frac{1}{2}$	$\frac{3}{8}$	,,	$\frac{5}{8}$
5	×	$3\frac{1}{2}$	$\frac{1}{2}$	,,	$\frac{5}{8}$	$2\frac{1}{2}$	×	$2\frac{1}{2}$ 2	$\frac{1}{4}$	,,	$\frac{1}{2}$
5	×	3	$\frac{7}{16}$	to	$\frac{5}{8}$	$2\frac{1}{4}$	×		$\frac{3}{16}$	,,	$\frac{1}{2}$
5	×	$2\frac{1}{2}$	$\frac{5}{16}$	,,	$\frac{1}{2}$	2	×		$\frac{3}{16}$	,,	$\frac{3}{8}$
$4\frac{1}{2}$	×	$4\frac{1}{2}$	$\frac{7}{16}$	,,	$\frac{5}{8}$	$1\frac{3}{4}$	×	$\frac{3}{4}$	$\frac{3}{16}$	,,	$\frac{3}{8}$
$4\frac{1}{2}$	×	4	$\frac{7}{16}$	,,	$\frac{5}{8}$	$1\frac{1}{2}$	×	$\frac{1}{2}$	$\frac{3}{16}$	,,	$\frac{3}{8}$
4	×	5	$\frac{1}{2}$	and	$\frac{5}{8}$	$1\frac{1}{4}$	×	$\frac{1}{4}$	$\frac{3}{16}$	,,	$\frac{1}{4}$
4	×	$4\frac{1}{2}$	$\frac{1}{2}$	,,	$\frac{5}{8}$	$1\frac{1}{8}$	×	$\frac{1}{8}$	$\frac{1}{8}$	and	$\frac{3}{16}$
4	×	4	$\frac{3}{8}$	to	$\frac{5}{8}$	$1\frac{1}{8}$	×		$\frac{1}{8}$	,,	$\frac{3}{16}$
4	×	$3\frac{3}{4}$	$\frac{3}{8}$	,,	$\frac{5}{8}$	1	×		$\frac{1}{8}$	,,	$\frac{3}{16}$
4	×	$3\frac{1}{2}$	$\frac{3}{8}$	,,	$\frac{5}{8}$	1	×	$\frac{7}{8}$	$\frac{1}{8}$	,,	$\frac{3}{16}$
4	×	3	$\frac{3}{8}$	,,	$\frac{5}{8}$	7	×	$\frac{7}{8}$	$\frac{1}{8}$	,,	$\frac{3}{16}$
4	×	$2\frac{3}{4}$	$\frac{3}{8}$	,,	$\frac{1}{2}$	$\frac{7}{8}$	×	$\frac{3}{4}$	$\frac{1}{8}$	,,	$\frac{3}{16}$

GRATE BARS.

Width.			Thickness.		Width.			Thickness.	
4	×	$\frac{7}{8}$	×	$\frac{3}{8}$	3	×	1	×	$\frac{1}{2}$
$3\frac{5}{8}$	×	$1\frac{3}{8}$	×	$\frac{7}{16}$	3	×	1	×	$\frac{1}{2}$
$3\frac{5}{8}$	×	$1\frac{1}{8}$	×	$\frac{1}{2}$	3	×	$\frac{3}{4}$	×	$\frac{3}{8}$
$3\frac{1}{2}$	×	$1\frac{1}{2}$	×	$\frac{1}{2}$	$2\frac{1}{2}$	×	1		$\frac{1}{2}$
$3\frac{1}{2}$	×	1	×	$\frac{1}{2}$					

RAILS.

Bridge	12	to	16	lbs.	per	yard.
,,	22	,,	24	,,		,,
T	14	,,	18	,,		,,
,,	24	,,	26	,,		,,
Train	22	,,	28	,,		,,
Street	36	,,	38	,,		,,

SASH IRON.

Wide.		Thick.	Circle.		
2	×	$\frac{3}{4}$	$1\frac{1}{2}$	×	$\frac{7}{16}$
$1\frac{3}{4}$	×	$\frac{3}{4}$	$1\frac{1}{4}$	×	$\frac{7}{16}$

GLUT IRON.

$3\frac{1}{2}$	×	$1\frac{3}{4}$
$1\frac{3}{4}$	×	$1\frac{1}{2}$

Advertisements.

CHANNEL IRON.

$$4 \times 2 \times \tfrac{5}{16}$$

ANGLES.

	Size.		Thickness.
	6	× 6	$\tfrac{1}{2}$ to 1 in.
	6	× 5	$\tfrac{1}{2}$,, 1 ,,
	6	× 4	$\tfrac{7}{16}$,, $\tfrac{7}{8}$,,
	6	× 3½	$\tfrac{7}{16}$,, 1 ,,
	6	× 3	$\tfrac{3}{8}$,, $\tfrac{7}{8}$,,
	5½	× 4½	$\tfrac{7}{16}$,, $\tfrac{7}{8}$,,
	5½	× 3½	$\tfrac{7}{16}$,, $\tfrac{7}{8}$,,
	5	× 5	$\tfrac{1}{2}$,, 1 ,,
	5	× 4½	$\tfrac{7}{16}$,, $\tfrac{7}{8}$,,
	5	× 4	$\tfrac{7}{16}$,, $\tfrac{7}{8}$,,
	5	× 3½	$\tfrac{3}{8}$,, $\tfrac{7}{8}$,,
	5	× 3	$\tfrac{5}{16}$,, $\tfrac{7}{8}$,,
	4½	× 4½	$\tfrac{7}{16}$,, $\tfrac{7}{8}$,,
Round back	4	× 4½	$\tfrac{3}{8}$,, $\tfrac{7}{8}$,,
	4	× 4	$\tfrac{3}{8}$,, $\tfrac{7}{8}$,,
Obtuse	4	× 4	$\tfrac{3}{8}$,, $\tfrac{5}{8}$,,
	4	× 3½	$\tfrac{5}{16}$,, $\tfrac{7}{8}$,,
	4	× 3	$\tfrac{3}{8}$,, $\tfrac{7}{8}$,,
Bulb	4	× 2½	$\tfrac{3}{8}$,, $\tfrac{7}{16}$,,
	4	× 4	$\tfrac{1}{4}$,, 1 ,,
	4	× 3½	$\tfrac{5}{16}$,, $\tfrac{7}{8}$,,
	4	×	$\tfrac{1}{4}$,, $\tfrac{3}{4}$,,
	4	× ½	$\tfrac{1}{4}$,, $\tfrac{5}{8}$,,
	3½	×	$\tfrac{1}{4}$,, $\tfrac{7}{8}$,,
	5½	×	,, 1 ,,
Round back	5	×	,, 1 ,,
	' 3½	×	$\tfrac{3}{8}$,, $\tfrac{5}{8}$,,
	3	×	$\tfrac{1}{4}$,, $\tfrac{3}{4}$,,
	3	×	$\tfrac{5}{16}$,, $\tfrac{5}{8}$,,
	3	×	$\tfrac{5}{16}$,, $\tfrac{5}{8}$,,

	Size.		Thickness.
Bulb	3½	× 2¼	$\tfrac{5}{16}$ to $\tfrac{7}{16}$ in.
Round back	3¾	× 3	$\tfrac{3}{8}$,, $\tfrac{3}{8}$,,
	1¼	×	$\tfrac{5}{16}$,, $\tfrac{5}{8}$,,
Boiler section	3¼	× ¾	$\tfrac{5}{16}$,, $\tfrac{5}{8}$,,
	3	×	,, ,,
	3	×	$\tfrac{1}{8}$,, $\tfrac{3}{8}$,,
	3	× 2¾	$\tfrac{1}{8}$,, $\tfrac{5}{8}$,,
	3	× 2½	$\tfrac{1}{4}$,, $\tfrac{5}{8}$,,
	3	×	$\tfrac{3}{16}$,, $\tfrac{1}{2}$,,
	¾	× ¾	$\tfrac{1}{4}$,, $\tfrac{1}{2}$,,
	¾	× ½	,, $\tfrac{1}{2}$,,
	½	× ½	$\tfrac{7}{16}$,, $\tfrac{1}{2}$,,
	2½	×	$\tfrac{3}{16}$,, $\tfrac{1}{2}$,,
	2¼	× 1½	$\tfrac{3}{16}$,, $\tfrac{7}{16}$,,
	1¼	×	$\tfrac{3}{16}$,, $\tfrac{1}{2}$,,
		×	$\tfrac{1}{8}$,, $\tfrac{1}{2}$,,
	1½	×	$\tfrac{1}{16}$,, $\tfrac{7}{16}$,,
Square edge	¾	× ⅜	$\tfrac{1}{4}$,, $\tfrac{1}{2}$,,
	⅝	× ⅜	$\tfrac{1}{4}$,, $\tfrac{3}{8}$,,
	½	× ½	$\tfrac{1}{8}$,, $\tfrac{3}{16}$,,
	¼	× ¼	$\tfrac{1}{8}$,, $\tfrac{5}{16}$,,
	⅛	× ⅛	$\tfrac{1}{8}$,, $\tfrac{1}{4}$,,
		×	$\tfrac{3}{16}$
	$\tfrac{7}{8}$	×	$\tfrac{1}{8}$
	$\tfrac{3}{4}$	×	$\tfrac{1}{8}$

THE

SNEDSHILL IRON COMPY

NEAR SHIFNAL, SHROPSHIRE.

Nearest Railway Station— OAKEN GATES.

MANUFACTURERS OF

PUDDLED AND CHARCOAL WIRE RODS,

BOILER PLATES,

ANGLE, CABLE, RIVET,

Horse Shoe & Merchant Iron,

PLATING BARS, &c.

OF, VARIOUS QUALITIES, FROM BEST SELECTED

LILLESHALL PIG IRON.

BRAND.

H. S.  BEST

SNEDSHILL

H.R.MARSDEN, SOHO FOUNDRY, LEEDS,

PATENTEE AND ONLY MAKER OF THE WELL-KNOWN

750   IN ALL PARTS   **OF THE**
NOW
IN USE   **WORLD.**

EXTRACTS FROM TESTIMONIALS.

'Your STONEBREAKER works admirably: it astonishes everybody.'

'I am exceedingly pleased with it; the alterations made are decidedly great improvements.'

'It is all I can desire.'

SAVES THE LABOUR OF 50 HANDS!'

30
FIRST-CLASS
GOLD AND
SILVER
MEDALS.

'Your MACHINE is a fascination; a wonder.'

'Has not cost a penny for repairs since we had it of you in 1867.'

Intending Buyers are Cautioned against Purchasing or using any Infringement of H. R. M.'S numerous PATENTS.

CHAS. RYLAND & SON—*continued.*

Agents for Messrs. Fletcher, Solly, and Urwick, the Osier Bed Iron Company, G. & R. Thomas, the Deepfield Iron Company, the Willingsworth Iron Company, the Executors of the late W. Mathews, Esq., South Staffordshire;

The Brymbo Coal and Iron Company, Wrexham, North Wales;

The Right Hon. Earl Granville, and Messrs. Stanier & Co., North Staffordshire;

The West Yorkshire Iron and Coal Company, Leeds, for their different makes of Pig Iron.

Also Agents for the Ulverston Mining Company, for their Hæmatite Ore, Ulverston;

The Wyken Colliery Company, near Coventry, for their Bedworth Balls Iron Stone; and Messrs. Geo. Skey & Co.'s Wilmecote Mine, near Tamworth; and also for Messrs. Thomas Adams & Co., and Mr. Robert Beswicks, North Staffordshire Mines; also Sole Agents in the South Staffordshire District for Messrs. the Llynvi, Tondu, and Ogmore Coal and Iron Company's well-known brand of Welsh Bars, South Wales.

PRESENT PRICES—*continued.*

			PER CWT.
Rounds, $\frac{5}{8}$ in. and upwards	... ...	24/
,, 9/16 and $\frac{1}{2}$ in. ...	... ...	26/
,, 7/16 and $\frac{3}{8}$ in. ...	... ...	28/
,, 5/16 in. ...	... ...	30/
,, $\frac{1}{4}$ in.	... ...	32/

RIVET IRON, same prices as above.
CHAIN IRON, £2 per Ton extra.
(B) Best **BARS** and **RODS,** extra per Cwt., 3/.

'BOWLING' BOILER PLATES.

Plates under $2\frac{1}{2}$ Cwt., each	... ...	29/
$2\frac{1}{2}$ Cwt. and under 3 Cwt.	... ... ·	30/
3 ,, ,, $3\frac{1}{2}$,,	... ...	32/
$3\frac{1}{2}$,, , 4 ,,	... ...	34/
4 ,, , 5 ,,	... ...,	37/
5 ,. ,, 6 ,,	... ...	40/
6 ,, ,, 7 ,,	... ...	43/
7 ,, and upwards	... ...	46/

Hammered and Chequered Plates; and all Plates differing from a square form or regular taper, extra per Cwt., 3/.
Plates exceeding 6 ft. 0 in. wide, 2/ per Cwt. extra.

SHEETS, 11 to 17, W.G.	... ...	31/
STRIPS for Welded Tubes, 1 to 10, W.G.	... ...	31/
Ditto ditto 11 to 14, do.	... ...	32/

'BOWLING' ANGLE IRON.

L & T Iron not exceeding 10 united inches ... ...
For each additional inch, extra per Cwt., 1/.

'BOWLING,' and upon the Thin Edge B*O,
RAILWAY TYRE BARS.

Under $3\frac{1}{2}$ Cwt.	... ...	25/6
$3\frac{1}{2}$ Cwt. and under 4 Cwt.	... ...	27/
4 ,, ,, $5\frac{1}{2}$,,	... ...	29/
$5\frac{1}{2}$,, and upwards ...	... ...	33/
Tyre Bars bent, extra per Ton ...	... ...	7/6

TYRES, WELDED and BLOCKED,
OR
'BOWLING' WELDLESS TYRES.

Under $3\frac{1}{2}$ Cwt.	... ...	26/6
$3\frac{1}{2}$ Cwt. and under 4 Cwt.	... ...	28/
4 ,, ,, $5\frac{1}{2}$,,	... ...	31/
$5\frac{1}{2}$,, and upwards ...	... ...	35/

'BOWLING' RAILWAY AXLES.

Swaged with Collars, or to Sketch, under 4 Cwt.	... ...	24/6
4 Cwt. and under 5 Cwt.	... ...	28/
5 ,, ,, 6 ,, ...	... ...	29/
6 ,, and upwards ...	... ...	31/6

PRESENT PRICES—*continued.*

'BOWLING' DOUBLE CRANK AXLES.
(For Locomotives.)

From the Forge	55/
Rough Turned	65/
Webs cut out	80/
Finished	110/

'BOWLING' PISTON RODS.

Swaged to any dimensions according to weight.

'BOWLING' USES.

Of all descriptions, according to weight and workmanship.

'BOWLING' RIVETS.

	PER CWT.
$\frac{7}{8}$, $\frac{3}{4}$, 11/16th in. diameter	29/
$\frac{5}{8}$ ... „	31/
9/16ths	34/
$\frac{1}{2}$	36/
7/16ths	40/

'BOWLING' HOOPING.

1$\frac{3}{8}$ in. broad and upwards	29/

Under this size same as small rods.

SASH IRON	27/
Slipers and Sock Bars	28/

'BOWLING' MOULDS.

Share, Hammer, and Arm	31/
Triangular	28/
Round and Skef Plates	31/

'BOWLING' WELDLESS HOOPS,

For strengthening Boiler Flues	45/

LONDON OFFICE—

114 CANNON STREET, E.C.—H. E. DRESSER, *Agent.*

UNITED STATES AGENT—

J. S. KENNEDY & CO., 41 Cedar Street, New York.

CANADA AGENT—

S. WADDELL & CO., 27 St. John Street, Montreal.

THE DARLASTON
STEEL & IRON COMPANY,
LIMITED,
WEDNESBURY.
10th JULY, 1873.

MANAGING DIRECTOR—WILSON LLOYD, ESQ.

PRICE LIST of BAR IRON, HOOP IRON, SHEET IRON, AND STEEL,

MANUFACTURED AT THE DARLASTON GREEN IRON WORKS, AND KING'S HILL IRON WORKS, WEDNESBURY, SOUTH STAFFORDSHIRE, ENGLAND.

This Price List is subject to alteration at any time without Notice.

Iron in bundles, namely—Sheets, Hoops, small Rounds and Squares, and Scrolls, will be delivered free alongside ship.—In London at 17/6 per ton extra; in Hull at 16/ per ton extra; in Liverpool at 12/6 per ton extra, in parcels of not less than 10 tons.

Iron not bundled, namely—Plates and Bars, will be delivered free alongside ship. In London at 15/ per ton extra; in Hull at 14/ per ton extra; in Liverpool at 10/ per ton extra, in parcels of not less than 10 tons.

TRADE MARK.

CHAMPION.

BAR IRON

Per Ton at Works.

	£	s.	d.
Flat, from 1 in. up to 6 in. wide by 4 in. and upwards	13	0	0
„ wider than 6 in., same as plates.			
„ thinner than ¼ in., same as hoops.			
Round or Square—			
from ¼ in. to 3 in.	13	10	0
over 3 in. to 3½ in.	13	10	0
over 3½ in. to 4 in.	14	0	0
over 4 in. to 4½ in.	15	0	0
over 4½ in. to 5 in.	15	10	0
Rounds only over 5 in. to 5¼ in.	16	10	0
Boiler Rivet Iron	14	10	0
7/16 in., round or square	13	5	0
⅜ in. ditto	13	15	0
5/16 and No. 1 and 2 W.G., ditto	14	5	0
¼ and No. 3 ditto	14	15	0
No. 4 ditto	15	0	0
No. 5 ditto	15	10	0
3/16 and No. 6 ditto	16	10	0
No. 7 ditto	17	10	0
5/32 and No. 8 ditto	19	0	0
No. 9 ditto	20	10	0
Best Bars and Rods, extra	1	0	0
Best Best do. do.	2	0	0

E 2

THE DARLASTON STEEL & IRON COMPANY—*continued.*

STEEL.

	£	s.	d.		£	s.	d.
Best Turning Tool Steel	3	0	0	Cotter and Drift Steel ..	1	16	0
Best Ingot Cast Steel ..	3	0	0	Best Cast Rolled Awl Blade ..	3	0	0
Best Cast Snap Steel ..	3	0	0	Best Tack Steel	1	14	0
Best Welding Cast Steel	3	0	0	Cast File Steel	2	6	0
Best Cast Tap and Die Steel ..	3	0	0	Rubber File Steel	1	16	0
Best Double Shear	3	0	0	Best Cast Gun Lock	3	0	0
Single Shear	2	4	0	Best Gun Lock	1	14	0
Blister Steel	1	14	0	Tilted Key Steel	1	5	0

The undermentioned Brands are the Trade Marks and Brands of the Company, and are stamped upon Iron and Steel at their Works, and are used to distinguish the various qualities of Iron and Steel made at the Works.

No. 1.—CHAMPION.

No. 2.—L   DARLASTON.

No. 3.—SAMUEL MILLS.

No. 4.—G. F.   G. F.

No. 5.—LLOYD'S DARLASTON.

No. 6.—LLOYD'S CHARCOAL.

No. 7.—L   STEEL.

No. 8.—   STEEL.

TRADE MARK.

CHAMPION.

SASH IRON AND MOULDINGS.

Secs. 47, 2, 312, 188, 3, 4, 5, 187, 13, 14, 15, 372, 19, 18, 210, 211, 212, 213, 214, 25, 26, 27, 28, 33, 34, 35, 184, 185, 247, 248, 249, 269, 267, 268, 1, 38, 39, 183, 37, 36, 46, 186, 45, 44, 282, 262, 284, 273, 279, 217, 283, 286, 280, 277, 285, 286, 340, 341, 344, 346, 345, 375, 343, 342, 373, 156, 381, 374, 298, 322, 338, 290, 56, 265, 263, 313, 315, 250 .. } £16 0 0

Secs. 6, 7, 8, 9, 10, 11, 12, 17, 16, 20, 42, 43, 215, 321, 21, 22, 23, 24, 29, 30, 31, 32, 314, 316, 317, 318, 240, 239, 319, 320, 272, 246, 245, 241, 242, 243, 244, 205, 206, 294, 207, 295, 208, 339, 209, 296, 270, 288, 48, 49, 57, 58, 310, 311, 337, 50, 51, 52, 53, 237, 311, 337, 50, 51, 52, 53, 237, 238, 297, 257, 54, 55, 40, 41, 189, 301, 303, 376, 377, 378, 379, 380 .. .. } 15 10 0

Secs. 217, 107, 324, 108, 292, 226, 325, 293, 106, 227, 425, 384, 385, 386, 261, 253, 254, 363, 349, 350, 89, 96, 95 .. .. } 15 10 0

Secs. 104, 329, 103, 328, 327, 102, 300, 224, 105, 274, 299, 161, 291 .. .. } 15 5 0

HALF ROUND, OVAL, AND HALF OVAL BARS.

⅞ in. and upwards, £13. 10s.; ¾ in., £14; ⅝ in., £14.! 10s.; ½ in., £15; ⅜ in., £15. 10s.; if 3/16 in. thick, extra 10s.; if ⅛ in. thick, extra £1.

ADDRESS—

THE DARLASTON STEEL & IRON COMPANY, LIMITED,

WEDNESBURY.

JOHN HORSLEY,

IRON AND METAL BROKER,

AND

COMMISSION AGENT.

ESTIMATES, Prices, Sections. and Drawings forwarded upon application for all descriptions of Machinery, Tools, Iron, Ironwork, and Colliery or Railway Materials.

SPECIFICATIONS AND DRAWINGS COPIED FOR ANY KIND OF WORK IN ANY PART OF THE COUNTRY,

St. Ann's Square,

MANCHESTER.

DESCRIPTION WITH SECTIONS

OF

I R O N

ROLLED BY

W. Barrows & Sons

BLOOMFIELD IRON WORKS,

TIPTON,

STAFFORDSHIRE.

MANUFACTURERS OF ALL KINDS AND SIZES OF

Merchant Iron, Bars, Hoops, Strips, Plates, Sheets, and
Tee Iron, &c. &c.

WILLIAM BARROWS & SONS—*continued.*

SIZES.

FLATS.

$\frac{1}{4}$ in. wide to	$\frac{3}{16}$ in. thick.			$2\frac{1}{2}$ in. wide to	$2\frac{1}{4}$ in. thick.		
$\frac{3}{8}$	,,	$\frac{5}{16}$	,,	$2\frac{3}{4}$	,,	$2\frac{1}{2}$	,,
$\frac{7}{16}$	,,	$\frac{3}{8}$	,,	3	,,	$2\frac{3}{4}$	,,
$\frac{1}{2}$	,,	$\frac{7}{16}$	,,	$3\frac{1}{4}$	,,	3	,,
$\frac{9}{16}$	,,	$\frac{1}{2}$	,,	$3\frac{1}{2}$	,,	$3\frac{1}{4}$	,,
$\frac{5}{8}$	,,	$\frac{9}{16}$	,,	$3\frac{3}{4}$	,,	$3\frac{1}{2}$	,,
$\frac{11}{16}$	,,	$\frac{5}{8}$	,,	4	,,	$3\frac{1}{2}$	,,
$\frac{3}{4}$	,,	$\frac{5}{8}$	,,	$4\frac{1}{4}$	,,	$3\frac{1}{2}$	,,
$\frac{7}{8}$	,,	$\frac{5}{8}$	,,	$4\frac{1}{2}$	,,	4	,,
1	,,	$\frac{7}{8}$	,,	$4\frac{3}{4}$	,,	4	,,
$1\frac{1}{8}$	,,	1	,,	5	,,	3	,,
$1\frac{1}{4}$	,,	1	,,	$5\frac{1}{4}$	,,	2	,,
$1\frac{3}{8}$	,,	$1\frac{1}{4}$	,,	$5\frac{1}{2}$	,,	2	,,
$1\frac{1}{2}$	,,	$1\frac{3}{8}$	,,	$5\frac{3}{4}$	,,	$2\frac{1}{2}$	,,
$1\frac{5}{8}$	,,	$1\frac{1}{2}$	,,	6	,,	$2\frac{1}{2}$	,,
$1\frac{3}{4}$	,,	$1\frac{1}{2}$	,,	$6\frac{1}{2}$	,,	$2\frac{1}{2}$	,,
2	,,	$1\frac{3}{4}$	,,	7	,,	$1\frac{1}{4}$	,,
$2\frac{1}{4}$	,,	2	,,	$7\frac{1}{2}$	,,	1	,,

ROUNDS.—$\frac{1}{8}$ in. to 4 in. diameter.

SQUARES.—$\frac{3}{16}$ in. to $3\frac{3}{4}$ in. ,,

OVALS.—$\frac{5}{16}$ in. to 1 in. wide; any thickness.

WILLIAM BARROWS & SONS—*continued.*

SIZES.

ANGLE IRON, Unequal Sided.

$1\frac{1}{4}$ in. $\times$ $\frac{3}{4}$ in.	$2\frac{1}{2}$ in. $\times$ 2 in.	$3\frac{1}{2}$ in. $\times$ 3 in.
$1\frac{1}{4}$,, $\times$ 1 ,,	3 ,, $\times$ $2\frac{1}{2}$,,	4 ,, $\times$ 3 ,,
$1\frac{1}{2}$,, $\times$ $1\frac{1}{4}$,,	$3\frac{1}{8}$,, $\times$ $1\frac{7}{8}$,,	4 ,, $\times$ $3\frac{1}{2}$,,
2 ,, $\times$ $1\frac{1}{2}$,,	$3\frac{1}{2}$,, $\times$ $2\frac{1}{2}$,,	

Sections 67 to 103.

ANGLE IRON, Obtuse.

$1\frac{1}{8}$ $\times$ $1\frac{1}{8}$	$2\frac{1}{2}$ $\times$ $2\frac{1}{2}$	$2\frac{3}{4}$ $\times$ $2\frac{3}{4}$

Sections 104, 105, 106.

TEE IRON.

1 in. $\times$ 1 in.	$1\frac{1}{4}$ in. $\times$ $1\frac{1}{4}$ in.	$1\frac{1}{2}$ in. $\times$ 2 in.
1 ,, $\times$ $1\frac{1}{2}$,,	$1\frac{1}{2}$,, $\times$ $1\frac{1}{2}$,,	(Beveled top)
$1\frac{1}{8}$,, $\times$ $1\frac{1}{2}$,,	$1\frac{1}{2}$,, $\times$ 2 ,,	$2\frac{1}{2}$ in. $\times$ $3\frac{1}{2}$ in.

Sections Nos. 107 to 117.

BEVELLED TYRE IRON.

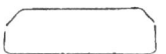

From $1\frac{1}{4}$ in. to 6 in. wide.   Sections Nos. 195 to 218.

CURVED TYRE and BOAT GUARD IRON.
Sections 187 to 194.

ROUND-EDGED FLATS.
$\frac{3}{4}$ in. to $2\frac{1}{4}$ in. wide.   Sections 219 to 228.

BEVELLED and STOCK HOOP IRON.
From 1 to $2\frac{1}{2}$ in. wide.   Sections 229 to 252.

FIRE BARS.
From 3 to 4 inches wide.   Sections 257 to 263.

SASH IRON.   Section 256.

TIP IRON.   Sections 253, 254, 255.

JOHN BAGNALL & SONS—*continued.*

THE FOLLOWING ARE THE PRESENT PRICES OF JOHN BAGNALL & SONS, LIMITED :—

	Per Ton
	£ s. d.
Bars—1 to 6 in. flat	
,, $\frac{1}{2}$ to 3 in. round and square	14 0 0
$\frac{1}{2}$, $\frac{5}{8}$, $\frac{3}{4}$ and $\frac{7}{8}$ flats.	
,, $3\frac{1}{8}$, $3\frac{1}{4}$, $3\frac{3}{8}$, and $3\frac{1}{2}$ in.	14 10 0
charged according to thickness.	
,, $3\frac{5}{8}$, $3\frac{3}{4}$, $3\frac{7}{8}$, and 4 in....	15 0 0
cut to exact lengths, 5s. per ton extra.	
,, $4\frac{1}{8}$, $4\frac{1}{4}$, $4\frac{3}{8}$, and $4\frac{1}{2}$ in. ...	16 0 0
,, $4\frac{5}{8}$, $4\frac{3}{4}$, and 5 in.	16 10 0
,, $5\frac{1}{4}$ and $5\frac{1}{2}$ in., round only	17 10 0
,, $5\frac{3}{4}$ and 6 in.	18 10 0
,, $6\frac{1}{4}$ and $6\frac{1}{2}$ in.	19 10 0
$\frac{7}{16}$, $\frac{3}{8}$, $\frac{5}{16}$, and $\frac{1}{4}$	
£14. 10s., £15 £15. 10s., and £16.	
,, $6\frac{3}{4}$ and 7 in.	20 10 0
,, $7\frac{1}{8}$ and $7\frac{1}{4}$ in.	21 10 0
,, 7, 8, and 9 in., flat	15 0 0
Turning Bars	14 0 0
Cable Bars	14 0 0
Plating Bars ... ---	14 10 0

Advertisements.

JOHN BAGNALL & SONS—*continued.*

	£	s.	d.
Best Bars	15	0	0
Horse Shoe Bars	15	10	0

Best Horse Shoe Bars, £1 per ton extra

	£	s.	d.
Best Rivet Iron...	15	10	0

Best Best Rivet Iron, £2. 10s.

	£	s.	d.
Best Boiler Strips	17	10	0

Hoops, 6 to 3¾ in. wide, not thinner than 14 W.G.	15		
,, 3½ to 2¼ in. wide, ,,	15	,,	15
,, 2 to 1⅝ in ,, ,,	17	,,	15
,, 1½ to 1⅜ in. ,, ,,	18		15
,, 1¼ to 1 in. ,, ,,	19	,,	15
⅞ in.	20	,,	16
¾ in.	20	,,	17
⅝ in.	20	,,	19
½ in. ,, ,,	20	,,	21

Best Hoops, £1 per ton extra.

Hoops thinner than the gauges mentioned to be charged extra 10s. per ton each gauge to 20 W.G., and 20s. per ton each gauge thinner than 20 W.G., cut to exact lengths 5s. per ton extra.

	£	s.	d.
Singles, to 20 in. W.G.	16	10	0
Doubles, 24 in.	18	0	0

JOHN BAGNALL & SONS—*continued.*

	£	s.	d.
Latten, 27 in.	19	10	0

£1 per ton extra for each gauge above

27 G.W.

	£	s.	d.
Boiler Plates, to 4 cwt.	16	10	0

Best Plates £1 per ton extra.

Best Best do., £2 per ton extra.

Plates above 4 cwt. and not exceeding

5 cwt., £1 per ton extra.

Do. 5 to 6 cwt., £2. 10s. per ton extra.

Do. 6 to $6\frac{1}{2}$ cwt., £3. 10s. per ton extra.

Do. $6\frac{1}{2}$ to 7 cwt., £4. 10s. per ton extra.

And £1 per ton extra for each $\frac{1}{2}$ cwt.

Plates 15 ft. long and above 4 feet wide,

£2 per ton each extra.

Cutting to irregular shapes, £1. 10s.

per ton extra.

	£	s.	d.
Angle Rails, 12 to 28 lbs. per yard ...			
Rails...			
Sash Iron			
Angle Bars, 1 to 4 in.	14	10	0
Best Angle Bars, $\frac{3}{4}$ to 4 in.	15	10	0
Gas Strip, 3 to 6 in. wide	13	5	0
$6\frac{3}{4}$ to $8\frac{1}{2}$ in.	13	15	0

AT OUR WORKS.

THE CHEQUE BANK—*continued.*

Head Office.

PALL MALL EAST, S.W.

City Office.

124 CANNON STREET, E.C.

Money can be paid to the credit of the CHEQUE BANK at its own Offices or at any of the following Bankers, where the funds of the Cheque Bank will be deposited :—

THE BANK OF ENGLAND.

THE WESTERN BRANCH OF THE BANK OF ENGLAND,
Burlington Gardens, Bond Street.

GLYN, MILLS, & CO.

WILLIAMS, DEACON, & CO.

NATIONAL PROVINCIAL BANK OF ENGLAND.

DIMSDALE, FOWLER, BARNARD, & CO.

CONSOLIDATED BANK, LIMITED.

ALEXANDERS, CUNLIFFES, & CO.

NATIONAL BANK OF SCOTLAND.

JAY, COOKE, McCULLOCH, & CO.

ALLIANCE BANK, LIMITED.

HERRIES, FARQUHAR, & CO.

R. TWINING & CO.

RANSOM, BOUVERIE, & CO.

CITY BANK.

NATIONAL BANK.

UNION BANK OF SCOTLAND.

MANCHESTER AND SALFORD BANK.

MANCHESTER AND COUNTY BANK.

Additions to this List will be published from time to time.

RYLANDS BROTHERS
LIMITED.

PRICES OF IRON WIRE, &c.

These Prices are obtained direct from the Firm.—*July* 10, 1873.

Best (ʀB/W) Iron Wire.

BRIGHT OR ANNEALED.

Per Bundle of 63lbs.	11s. 9d.	12s. 0d.	12s. 3d.	12s. 6d.	13s. 0d.	13s. 3d.	13s. 6d.
Nos.	0 to 6	7	8	9	10	11	12
Per Bundle of 63lbs.	13s. 9d.	14s. 3d.	14s. 9d.	15s. 6d.	16s. 3d.	17s. 0d.	17s. 9d.
Nos.	13	14	15	16	17	18	19
Per Bundle of 60lbs.	18s. 9d.	19s. 9d.	20s. 9d.				
Nos.	20	21	22				

Charcoal Wire .... 6s. per Bundle extra | Coppered Wire.......... 9d. per Bundle extra.
Tinned Wire .... 6s. ,, ,, | Wire drawn to pattern.. 3d. ,, ,,
Tinded Charcoal Wire.... 12s. ,, ,, | Charcoal Half-round Wire 10s. ,, ,,

Best Selected (ʀB/W) Spring Wire.

Per Bundle of 63lbs.	12s. 3d.	12s. 6d.	12s. 9d.	12s. 9d.	12s. 9d.	13s. 0d.	13s. 6d.
Nos.	0 to 6	7	7½	8	8½	9	10

Charcoal Spring Wire .... 6s. per Bundle extra.
Coppered Spring Wire .... 9d. ,, ,,

Best Best Prepared Bright (ʀB/W) Fencing Wire.

Nos. 0 to 6 .. £19 15 0 per Ton.
7 to 8 .... 20 15 0 ,,

Best Best Annealed Drawn (ʀB/W) Fencing Wire.

Nos. 0 to 6................ £18 5 0 per Ton. | No. 10 .... £21 5 0 per Ton.
7 .... 19 0 0 ,, | 11 .... 21 15 0 ,,
8 .... 19 15 0 ,, | 12 .... 22 5 0 ,,
9 .... 20 5 0 ,, |
Dipping in boiling oil, 5s. per ton extra.

Best Best Drawn Galvanized (ʀB/W) Fencing Wire.

Nos. 1 to 6 .... £22 0 0 per Ton. | No. 9 .... £24 15 0 per Ton.
7 .. .. 23 0 0 ,, | 10 .... 25 15 0 ,,
8 ....:............ 23 15 0 ,, |

Rolled | W I W | Fencing Wire.

BLACK. | GALVANIZED.
Nos. 1 to 4 .... £15 10 0 per Ton. | Nos. 1 to 4 .... £19 15 0 per Ton.
5 .... 16 0 0 ,, | 5 .... 20 5 0 ,,
6 .... 16 15 0 ,, | 6 .... 21 0 0 ,,

Best Best Galvanized (ʀB/W) Fencing Strand, 7 Ply.

Per Cwt.	27s. 9d.	28s.	28s. 6d.	29s.	30s. 6d.	31s. 3d.	32s. 3d.	34s.	35s. 9d.	36s. 6d.	38s.
Nos.	0	1	2	3	4	5	6	7	8	9	10

All Sizes and Plies quoted for on application.

RYLANDS BROTHERS, LIMITED—*continued.*

PRICES OF IRON WIRE, &c.

Machine-Cut Fencing Staples.

Per Cwt.	....	22s. 6d.	23s. 0d.	23s. 6d.
Nos.	.. 5 and 6	7	8	

Galvanized Staples, 7s. 6d. per cwt. extra.　　STRAINING SCREWS, 15s. each.
Iron Kegs to hold 1 cwt. of Staples, 2s. 3d. each.

Best Best Drawn Killed Galvanized Telegraph Wire.

(*Joined in Half-mile Lengths to No. 9 inclusive with Rylands' Patent Joint.*)

Nos. 0 to 6	....	£24 5 0 per Ton.	No. 10	....	£26 15 0 per Ton.
7 & 8	....	25 5 0 ,,	11	....	27 5 0 ,,
9	....	26 0 0 ,,	12	....	28 5 0 ,,

Best Refined Telegraph Wire, £3 per ton extra.
Charcoal Telegraph Wire, £10. 10s. per ton extra.

Best Improved Galvanised Wire.

Per Bundle of 63 lbs.	14s.	14s. 3d.	14s. 6d.	15s.	15s. 6d.	15s. 9d.	16s.
Nos.	0 to 6	7	8	9	10	11	12
Per Bundle of 63 lbs.	16s. 3d.	16s. 9d.	17s. 9d.	18s. 6d.	19s. 9d.	20s. 6d.	21s. 3d.
Nos.	13	14	15	16	17	18	19
Per Dozen lbs.	4s. 10d.	5s.	5s. 3d.	5s. 9d.	6s.	6s. 6d.	7s. 6d.	8s. 3d.	9s.
Nos.	20	21	22	23	24	25	26	27	28

Annealed Tinned Iron Wire.

Per Dozen lbs.	5s.	5s. 3d.	5s. 6d.	5s. 9d.	6s.	6s. 6d.	7s.	7s. 6d.
Nos.	18	19	20	21	22	23	24	25
Per Dozen lbs.	8s. 3d.	9s. 9d.	10s. 9d.					
Nos.	26	27	28.					

Best Weaving and Binding Wire.

Per Dozen lbs.		4s. 4d.	4s. 7d.	4s. 10d.	5s.	6s. 5d.	6s. 8d.	6s. 11d.	7s. 2d.
Nos.		23	24	25	26	27	28	29	30
Per Dozen lbs.	7s. 5d.	7s. 11d.	8s. 8d.	9s. 8d.	10s. 11d.	12s. 5d.	13s. 10d.	15s. 10d.	
Nos.	31	32	33	34	35	36	36½	37	

Charcoal Wire to No. 26, 1s. 1d. per dozen extra.
Bottling Wire cut in lengths, 6d. per dozen extra.　Thicker Wire 1s. 6d. per bundle extra.
Wire wound in ½ and 1 lbs. 1s. 6d. per bundle extra ; ¼ lbs. 2s. per bundle extra.
Wire wound in 1 and 2 oz. hanks, 1d. per lb. extra.
Dudley Bagging, 3d., Cotton Bagging, 1¾d. per bundle extra.
Papering, 1d. per bundle extra.

Delivered Free in Liverpool or Manchester.

If in London, 15s per ton extra, or for 5 ton lots 12s. 6d. extra, f.o.b.
If in Edinburgh, 26s. 8d. per ton extra, or for 2 ton lots 18s. 4d. extra.
If in Dublin *via* Liverpool, 15s. per ton extra.
If in Dublin *via* Holyhead, 18s. 4d. per ton extra.
If in Glasgow :—
　　By Steamer, 10s. 10d. per ton extra.
　　By Rail, 21s. 8d. per ton extra, or for 2 ton lots 15s. extra.
If in Hull, 20s per ton extra, or for 2 ton lots 10s. extra.

Terms of Payment.

2½ per cent. for cash on 10th of month following delivery.

GUEST BROTHERS & CO.

(ESTABLISHED 1854),

MANUFACTURERS OF

CHILLED AND GRAIN

ROLLS,

TOOTH WHEELS

SHAFTING, & MILL & FORGE

Castings in General.

ROLLS TURNED TO ANY SECTION AND
FOR ANY PURPOSE.

TRAINS OF ROLLS FITTED UP COMPLETE.

VICTORIA IRON FOUNDRY,

WEST BROMWICH.

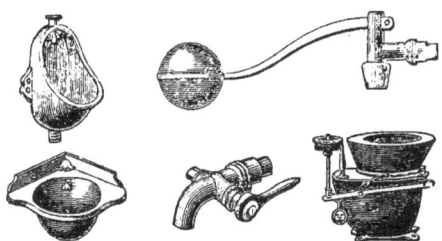
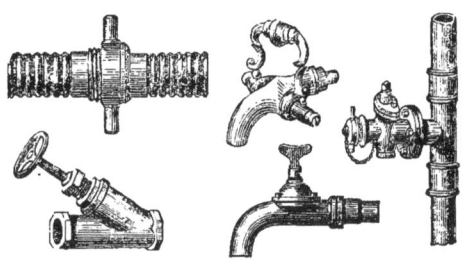
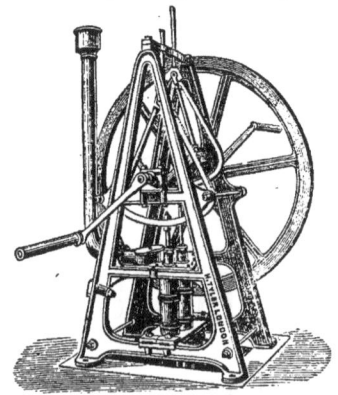
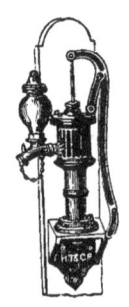

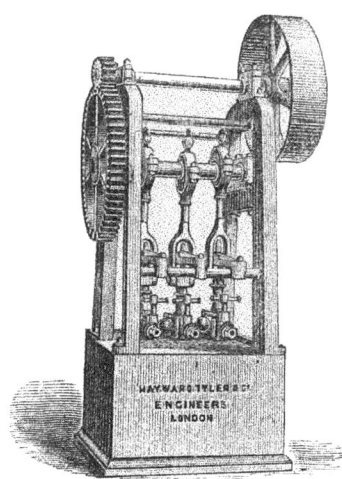
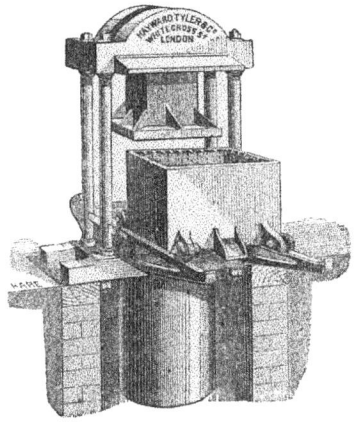

HAYWARD TYLER & Co.

(ESTABLISHED 1815),

THE OLDEST MAKERS IN THE KINGDOM OF

SODA-WATER MACHINERY

IN ALL ITS BRANCHES,

84 AND 85 UPPER WHITECROSS STREET,

LONDON.

PUMPS GAS GENERATOR, AND BOTTLING MACHINE

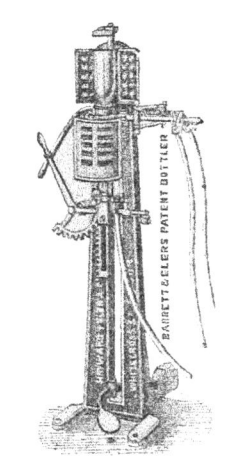

SODA-WATER MACHINE.

PATENT BOTTLING MACHINE.

LONDON, 1862.

VIENNA, 1873.

PARIS, 1867.

THE

PATENT
NUT & BOLT CO.

LIMITED,

THE PATENT NUT AND BOLT CO.
Continued.

MANUFACTURERS OF ALL KINDS OF

RAILWAY FASTENINGS,

INCLUDING

FISH PLATES,

SOLE PLATES, FISH BOLTS,

FANG BOLTS,

AND

STRAP BOLTS,

CUP HEAD AND DOG HEAD

SPIKES.

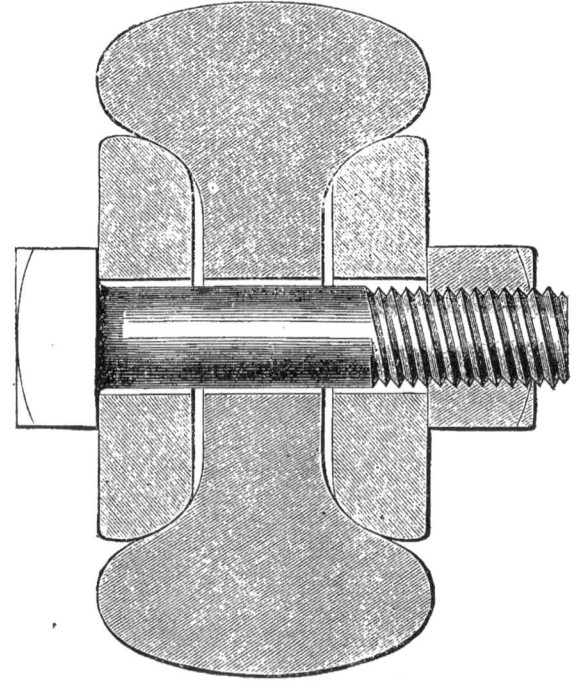

THE PATENT NUT AND BOLT CO.
Continued.

WORKS:

London Works, Birmingham.

Stour Valley Works, West Bromwich.

Midland Works, Soho, near Birmingham.

Cwm Bran Works, Blast Furnaces

Collieries, and Rolling Mills,

near Newport, Monmouthshire.

AGENCIES:

24 Budge Row, Cannon Street, London.

41 The Temple, Dale Street, Liverpool.

20 Mansfield Chambers, St. Ann's Square, Manchester.

40 Sandhill, Newcastle-on-Tyne.

10 Norfolk Street, Sheffield.

147 Trongate, Glasgow.

ADDRESS—

CHIEF OFFICES, LONDON WORKS,

NEAR BIRMINGHAM.

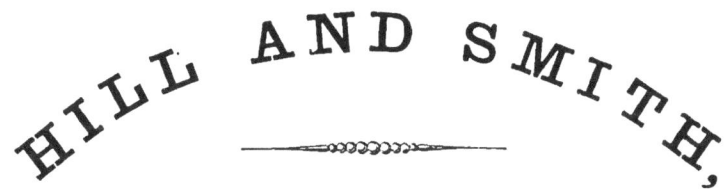

IRON WORKS, BRIERLY HILL,

MANUFACTURE

WROUGHT IRON SHAFTS,

CRANK SHAFTS FOR MARINE ENGINES,

PISTON RODS, CONNECTING RODS,

ALL USE IRON OF THE LARGEST WEIGHTS AND SIZES.

MANUFACTURERS BY APPOINTMENT

OF

GRIFFITHS' PATENT PUDDLING MACHINE.

HURDLES AND IRON FENCING OF ALL KINDS,

INCLUDING

WIRE FENCING FOR PARKS AND ESTATES.

PRICES FOR FENCING PARKS AND ESTATES WILL BE
FORWARDED TO NOBLEMEN AND OTHERS TO ANY PART
OF THE UNITED KINGDOM ON APPLICATION.

The "Sicker"
Double Grip-Bolt Safes. Strong.
Room Doors &c. for Bankers,
Jewellers. Merchants &c
Fire, Wedge, Drill & Gunpowder-proof

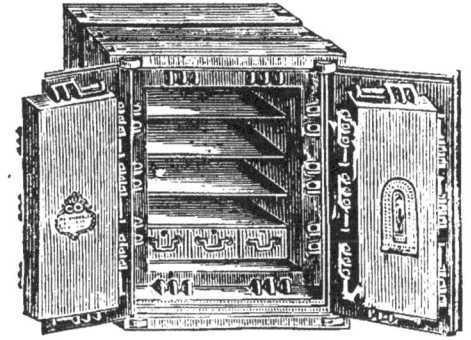

The only Safes manufactured
which effectually resist Fire.
and the skilled Burglar
Inventor, Patentee & Sole manufacturer.
James Felton Elwell,
The "Vulcan" Works, Birmingham.

THOMAS L. NICKLIN,

CROWN IRON WORKS

SMETHWICK,

MANUFACTURES

BARS, ANGLE IRON

SMALL ROUNDS AND SQUARES,

PLATING BARS, AND RIVET IRON,

BEST AND BEST BEST BEST,

IRON OF ALL KINDS.

POSTAL ADDRESS,

SMETHWICK.

All trains stop at SMETHWICK STATION, which is distant only five minutes' walk from the Works.

JOSEPH PEARSON,

MAKES AT

WINDMILL END FURNACES

Nos. 1, 2, and 3

MELTING PIG IRON,

Strong Forge Mine Iron,

·AND

OTHER KINDS OF PIG IRON.

The Earl of Dudley's THICK COAL
is used.

G. & W. UNDERHILL—*continued.*

,INCLUDING

S. C.

EARL OF DUDLEY'S ROUND OAK BARS

B. B. H.

THE MITRE.

G. B. THORNEYCROFT & CO.

"THE LION,"   "JOHN BAGNALL & SONS,"
"CHILLINGTON RODS,"   "MONMOOR BARS AND PLATES,"
"WILDEN SHEETS,"   "SNEDSHILL WIRE RODS & PLATES,"

Adams', Regent, and Pelsall Bars, Sheets, and Hoops, and most other descriptions of Iron.

MANUFACTURERS SUPPLIED WITH

SPELTER,
LEAD, AND
COPPER.

POSTAL ADDRESS:—

WOLVERHAMPTON.

H

JOHN DAWES & SONS—*continued.*

Small Rounds and Squares,

ANGLES, AND ALL KINDS OF BEST IRON SHEETS & BOILER PLATES,

Plating Bars, Small Angles,

AND

FANCY IRON.

Postal Address,

JOHN DAWES & SONS,
BROMFORD IRON WORKS,
WEST BROMWICH,
STAFFORDSHIRE.

H 2

S. GRIFFITHS & CO.—*continued.*

BARS, BOILER PLATES,

SHIP PLATES,

Sheet Iron of all Kinds,

INCLUDING

RUSSIAN SHEETS,
BLACK PLATE & CANADA PLATES;

ANGLES, T IRON

NAIL RODS, WIRE RODS,

FENCING RODS AND ALL KINDS OF WIRE;

SCOTCH PIG IRON

AND

ALL ENGLISH BRANDS OF PIG IRON.

Sales effected of all kinds of Iron Ore and Iron Stone.

Shippers of Iron to all parts of the World.

THE
CHILLINGTON COMPANY

Manufacture at their various Iron Works

Chillington Slit Nail Rods,

Bars, Bale-Hoops,

TRAM-RAILS, ANGLES,

BOILER PLATES,

SHEET IRON OF ALL KINDS.

Puddled Steel

AND

PIG IRON.

Brands:

 CHILLINGTON.

C C   L B

DAVID ROSE—*continued.*

GALVANISING.

SUGAR MOULDS FOR TINNING,

PAN AND TANK PLATES,

ALL KINDS OF BARS,

Small Rounds and Squares and Fancy Iron.

ALSO,

GALVANISING SHEETS

made at the Works, where both Galvanising and Corrugating
are carried on on a large scale.

PIG IRON

is also made at the Blast Furnaces, and

FIRE BRICKS

for Sale in the district.

POSTAL ADDRESS:—

DAVID ROSE,

MOXLEY.

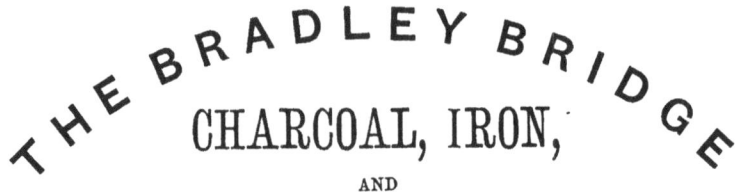

DISTRICT

Iron and Steel Company,

SMETHWICK, near BIRMINGHAM,

MANUFACTURE, AT THE

District and Dudley Port Iron Works,

MERCHANT BARS, HOOPS,

SHEET IRON, STRIPS, ANGLES,

BOILER PLATES,

AND

SMALL ROUNDS AND SQUARES,

ROOFING SHEETS, TANK PLATES, AND GALVANIZING SHEETS

TO SPECIFICATION.

TRADE MARK   **B. K. H.**
                          B. K. H.

Smethwick Railway Station a few minutes only from the Works.

Postal Address,

SMETHWICK, near BIRMINGHAM.

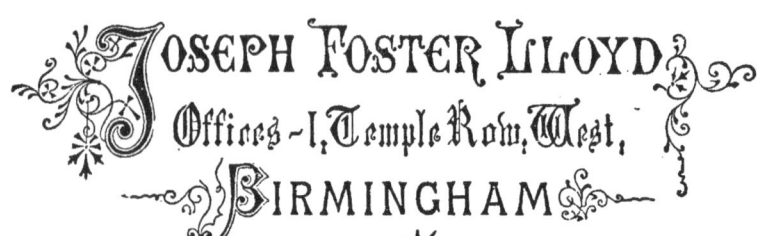

JOSEPH FOSTER LLOYD
Offices—1, Temple Row, West,
BIRMINGHAM

LABOR VINCIT OMNIA

DEPÔTS,

MONUMENT LANE STATION

RAILWAY WHARF,

CRESCENT BIRMINGHAM

CANAL WHARF

PIG IRON

COKE AND GENERAL

IRON,

METAL & MINERAL

MERCHANT,

BRANDS AND QUALITIES OF IRON.

BEST BEST.
TREBLE BEST.
CHARCOAL.

NE PLUS ULTRA.
H. L. M.
COLD BLAST.

VULTURE

PIG IRON.—Cold Blast, Hot Blast of the Best Brands of Staffordshire, Wales, and Yorkshire, supplied for Forges and Foundries, also Cold Blast Charcoal, and Hematite Pigs, for Malleable Foundries.

STEEL.—Every description of Rolled and Tilted Steel, both Cast and Blistered, made from Best Swedish Bars.

BAR, HOOP, STRIP, ANGLE, & TEE IRON of all sizes, to order.

BOILER, SHIP, & BRIDGE PLATES of all sizes, to order.

SHEET IRON, of the Best Brands, for Stamping, or for Galvanizing and Roofing purposes.

IRON & STEEL RAILS, WHEELS, & AXLES

RAILWAY SPIKES, BOLTS, NUTS, & RAILWAY MATERIALS of all descriptions.

IRON ORES imported, and all descriptions of Ore supplied.

COKES.—Yorkshire, Welsh, and Durham Cokes, for Foundries and Blast Furnaces.

ANNEALING PANS & ANNEALING ORE.

ENQUIRIES AND ORDERS SOLICITED AT THE ABOVE ADDRESS.

Lightning Source UK Ltd.
Milton Keynes UK
UKOW06f1945030116

265721UK00006B/510/P

9 781331 518266